CCTV for Wildlife Monitoring

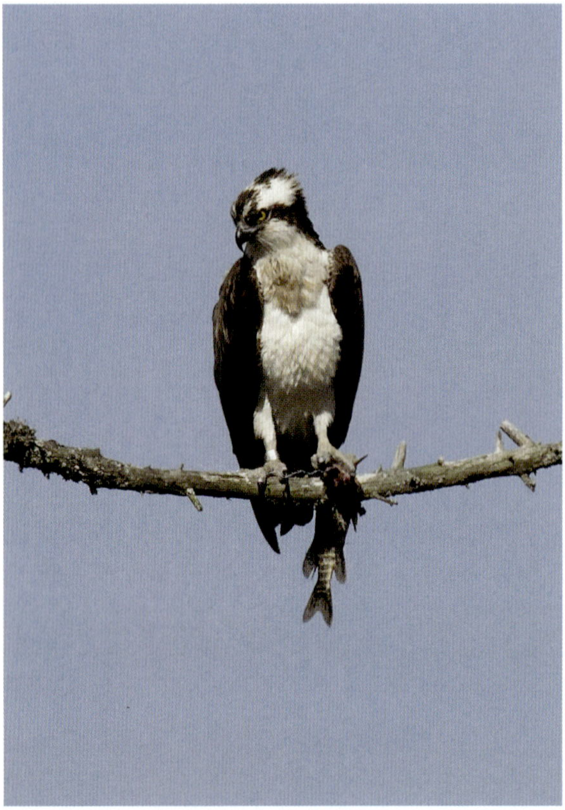

The osprey (*Pandion haliaetus*) is often monitored with CCTV

CCTV for Wildlife Monitoring

An Introduction

Susan Young

DATA IN THE WILD

Pelagic Publishing | www.pelagicpublishing.com

Published by Pelagic Publishing
www.pelagicpublishing.com
PO Box 725, Exeter EX1 9QU, UK

CCTV for Wildlife Monitoring: An Introduction
ISBN 978-1-78427-097-1 (Pbk)
ISBN 978-1-78427-096-4 (Hbk)
ISBN 978-1-78427-098-8 (ePub)
ISBN 978-1-78427-099-5 (Mobi)
ISBN 978-1-78427-100-8 (PDF)

This book should be cited as Young, S. (2016) *CCTV for Wildlife Monitoring: An Introduction*. Exeter: Pelagic Publishing, UK.

British Library Cataloguing in Publication Data
A catalogue record for this book is available from the British Library.

All images were produced by the author.

Cover images: Top left: Long-eared bat (*Plecotus* sp.) video still from a 1,000 TVL analogue CCTV camera. Top right: Badger (*Meles meles*) video still from an IP CCTV camera. Bottom left: Bank vole (*Clethrionomys glareolus*) video still from an IP CCTV camera. Bottom right: Blue tit (*Parus caeruleus*) and great tit (*Parus major*) video stills from an HD-TVI camera.

Contents

Preface

Ecology and the natural sciences have not tended to be technology-oriented subjects, as many monitoring techniques depend on human skills. However, technology not only follows from a need, it can also suggest applications, as the use of camera traps has shown, with an increasing number of researchers realising their value.

At present, wildlife CCTV is mainly used by large organisations such as wildlife trusts, wildlife parks and gardens, and is generally installed by CCTV technicians. However much of the technology and many of the applications do not require as much technical knowledge as might at first appear, and I believe that usage will only develop if it can be implemented by CCTV 'non-experts'.

The applications described in this book are based in the UK. In other countries distance and climate may be more challenging, as may the nature of the wildlife subjects. Equipment will need greater protection if monitoring potentially destructive wildlife, or if carried out in extreme climates. However, CCTV principles will be the same wherever they are used, and the great variety of wildlife found in countries other than the UK will supply a huge range of subjects to study.

CCTV is, I believe, the next logical step in wildlife monitoring, and in its various forms can enrich the experience of the wildlife enthusiast and increase the repertoire of the ecology consultant and the researcher

Susan Young
March 2016

Acknowledgements

Thanks to Albert Knott, Natural England Reserves Manager, Dartmoor NNRs and David Rickwood, Woodland Trust Devon Site Manager, for their encouragement; to Andrew Carr for his help with bats and Freya Womersley for her enthusiasm with fish; and last but not least to Sherpa Jim for everything.

List of video references

These short videos based on the material in each chapter can be found on YouTube (the URL is indicated in the figure caption by 🎞).

https://youtu.be/iXeLk76tR68	Chapter 1: Bird box nestlings.
https://youtu.be/_byGfYuYNBM	Chapter 2: Fox taken with a standard analogue camera.
https://youtu.be/j_yiNp-28W8	Chapter 2: Two badgers taken with a standard analogue camera.
https://youtu.be/U6bCPkRpG70	Chapter 2: Bat taken with a high-resolution analogue camera.
https://youtu.be/4AasL5VZJ00	Chapter 3: Fox cub taken with an HD-TVI camera.
https://youtu.be/4uNsy04XtZ8	Chapter 4: Roe deer taken with an IP camera.
https://youtu.be/rlpcfcHmB4Y	Chapter 4: Shrew in a mammal feeding station.
https://youtu.be/TavyLIOZ9Ek	Chapter 6: Enhancing low contrast.
https://youtu.be/c1NGiHsa5lE	Chapter 6: Video enhancement.
https://youtu.be/ZEzmoJkDnxQ	Chapter 7: Squirrel showing what can go wrong with video.
https://youtu.be/IgLjBtTKdMg	Chapter 7: Vole in a mammal feeding station.
https://youtu.be/C-WQ8Dd44Xw	Chapter 8: Pond dipping.
https://youtu.be/ul8Tf-eTTqA	Chapter 8: Frog catcher.
https://youtu.be/DW-zLZoIkPg	Chapter 8: Heron and fox.
https://youtu.be/3LNERsfi6Kc	Chapter 8: Tawny owl frog catcher.
https://youtu.be/HVHqrQ7NYN8	Chapter 8: Rock pool tiny fish.
https://youtu.be/O_XCc3i6R9g	Chapter 8: Rock pool sea snails.
https://youtu.be/DVZxoM5Wm24	Chapter 8: Rock pool predators.
https://youtu.be/BVpalb5nKh4	Chapter 8: Sleepy badger with cub.
https://youtu.be/GZNvRe4mGac	Chapter 8: Bird bath action.
https://youtu.be/wg2SUh-ohb8	Chapter 8: Tawny owls courting.
https://youtu.be/qWYZ3ojdhKQ	Chapter 9: Fish in a Dartmoor stream.
https://youtu.be/oEok3ESPCcw	Chapter 9: Fish in a deep pool.
https://youtu.be/GaHDKdbO508	Chapter 9: Larger fish.
https://youtu.be/6i4DUAAQ9T4	Chapter 9: Two barbastelle bats.
https://youtu.be/O_0whqdqCqw	Chapter 9: Barbastelle bat.
https://youtu.be/RRaAAMa4Qfg	Chapter 9: Long-eared bat.
https://youtu.be/PTjwV9pBvqY	Chapter 9: Unexplained bat behaviour.

Abbreviations

3G Third-generation mobile phone standard which allows transmission of email and the ability to browse the internet. Data speeds are variable but of the order of 1 megabit per second (Mbps) according to Ofcom, equivalent to a slow broadband. Some CCTV recorders and trail cameras can send images by 3G mobile technology.

4G Fourth-generation mobile phone standard, which is 5–7 times faster than 3G. According to Ofcom the average data transfer rate is 6 Mbps. It should be possible to transmit videos from CCTV recorders with 4G.

Ah Ampere hour. This describes the capacity of a battery in terms of the current drawn from it and the length of time that current can be maintained. Thus 8 Ah means a current of 8 amperes (amps) can be supplied for 1 hour, 4 amps for 2 hours, 1 amp for 8 hours and so on.

AV Audiovisual. This refers to the signal from a CCTV camera. Although not all CCTV cameras supply an audio signal, this notation is still used when describing inputs and outputs from recorders.

BNC A type of connector commonly used with CCTV cameras. BNC stands for Bayonet Neill–Concelman, the inventors, although it is sometimes described as a British Naval Connector.

Cat5 This is short for Category 5 and describes the standards for a twisted-pair cable type commonly used for computer networks, often called ethernet cable.

Cat6 This is a higher standard than Cat5 and offers greater waterproofing and electrical noise reduction.

DC Direct current. The current from batteries is DC as it flows in one direction. Many CCTV cameras and recorders use electrical adapters to convert mains alternating current (AC) to DC. In the UK, mains voltage is 240 V AC and most CCTV cameras and portable recorders require 12 V DC.

DSLR Digital single-lens reflex camera. The more expensive form of digital camera which also usually has detachable lenses.

DVR Digital video recorder. The recorders used for CCTV record in a digital format to a hard disc, SD card or similar digital storage device

EPDM A hard-wearing synthetic rubber material commonly used for pond liners or roofing.

GB Gigabytes. This describes the digital storage capacity of devices such as hard discs and means one thousand million bytes of data or 1,000 megabytes (MB).

GHz Gigahertz or one thousand million hertz. This refers to the frequency of the electromagnetic waves produced by CCTV wireless transmitters.

HDMI High-definition multimedia interface. This refers to a particularly high-quality audiovisual signal.

HD-TVI High-definition transport video Interface. This describes the latest form of high-definition analogue CCTV cameras that give a particularly high-quality signal.

IP Internet Protocol. This type of CCTV camera produces a digital signal and transmits it through the internet.

IP68 This is the International Protection Marking code for devices that are dust- and waterproof. The IP used here is not related to the IP used for Internet Protocol cameras.

LED Light-emitting diode. An electronic component that emits light when a voltage is applied to it. LEDs draw a low current and produce very little heat compared to other electrical light sources.

Mbps Megabits per second. This describes the speed at which digital data is transmitted.

nm Nanometres, or 10^{-9} m. This is the unit used to describe the wavelength of light.

PAL Phase alternating line. This describes the standard used to produce images on an analogue television and is relevant to CCTV videos. See Figure 2.3.

PIR Passive infrared sensor. This detects heat in the form of infrared waves. It is used in trail cameras and some CCTV cameras to detection motion. It measures the change in heat between a warm object and cooler surroundings.

POE Power over the ethernet. A system that allows power to be carried along with the video signal in an ethernet cable.

RCA Radio Corporation of America connector. Sometimes called a phono connector. This describes the type of connectors used in lower-quality CCTV systems. See Figures 2.10 and 2.11.

RG59 A standard of coaxial cable used to carry video signals. See Figure 2.6 for a diagram of the structure of an RG59 cable.

SCART This is a system using connectors with 12 pins (or holes). It once commonly connected televisions to recorders, although with the advent of HD televisions has been largely replaced by HDMI connectors. The letters are from Syndicat des Constructeurs des Appareils Radiorécepteurs et Téléviseurs, the French designers.

SD Secure digital. A small card for storing digital data commonly used in portable DVRs and digital cameras.

TVL Television lines. This gives a measure of the resolution of analogue CCTV cameras and describes the maximum number of lines that can be seen individually. See Figure 2.4.

VGA Video graphics array. This is a 15-pin connector commonly used on recorders and televisions to transfer video signals.

CCTV for wildlife monitoring – an overview

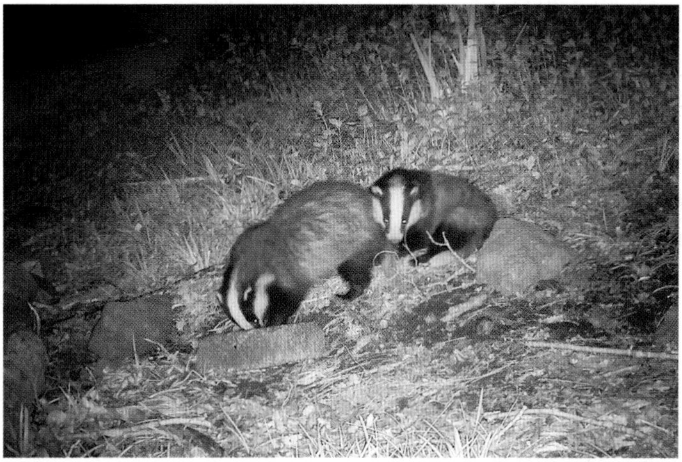

Badger (*Meles meles*) mother and cub (video screenshot from a high-resolution CCTV camera).

CCTV has great potential for monitoring wildlife, whether it is used to watch nesting birds in a bird box, or to unlock puzzles of animal behaviour for research. This book gives details of how to use CCTV for these applications, plus many others.

CCTV is not new, and television wildlife programmes often show examples of CCTV videos. Television applications generally work with large budgets and teams of people. They often use the most up-to-date equipment, and employ experts to solve technical difficulties.

However, this book is not about expertise and high-cost equipment; instead it aims to show how straightforward, low-cost solutions can be implemented by a novice to the subject.

The work is based on my experience using CCTV in this way, together with some pioneering research showing bat behaviour I believe has not been filmed before, and I am passionate about what I think is a powerful and useful tool. However, the use of this technology in wildlife monitoring can only develop when in the hands of the non-expert, as has been shown with trail cameras. I cover a range of applications, but there are many hundreds of exciting possibilities out there for those with a little knowledge.

Why CCTV?

There are many situations where trail cameras will do the job and the extra complexity of CCTV would not be required. However, CCTV should be used:

- When live view to a monitor is required (with mains power available).
- When live streaming via the internet is required (with mains power and an internet router available).
- For fast-moving wildlife such as bats, or cold animals such as wet otters or fish.
- When submersible cameras are required for more than a few hours.
- When flexible recording settings are required with:
  - Motion-triggered recording.
  - Pre-record.
  - Adjustable post-record times.
  - Masking of unwanted moving objects such as foliage.
  - Adjustable video quality, for example using low resolution for live streaming.
- When a variable-focus lens is required.
- When high-resolution video is required (either using IP cameras or HD-TVI).
- When high-quality weatherproof cameras are required.
- When several cameras are used at once, and a recorder with multiple inputs is available.
- When continuous filming is required.
- When flexible scheduling is required such as motion detection within certain time spans.
- When sophisticated cameras are required, such as those that prevent overexposed images when close to the camera, and that have in-camera processing to reduce noise and enhance the image.

How the book is structured

The book is intended for a wide range of interests and not everyone will wish to read every chapter at once. Jumping straight into the technology is rather daunting for the novice to the subject, and pages of detail on connectors and CCTV parameters, although necessary at some stage, will not always inspire enthusiasm. For this reason, each new section, where appropriate, has a Quick Start beginning, which will allow initial use of the equipment without the requirement to understand a great deal. Once experience has been gained in this way, it will be time to delve into the details.

Chapter guide

Chapters 1, 3, 8 and 11 are suitable for the novice or those who do not wish to delve into the technical detail too much.

- Chapter 1 is an introduction to CCTV and includes a Quick Start section.
- Chapter 3 is an introduction to HD analogue CCTV (HD-TVI), and has a Quick Start section. This is the best CCTV method if you have access to a mains supply.
- Chapter 8 has a host of examples of CCTV use for the wildlife enthusiast to inspire you.
- Chapter 11 compares trail cameras to portable CCTV.

Chapters 4, 6, 7 and 10 are for those who wish a little more detail of equipment and processes.

- Chapter 4 gives details of IP CCTV, which requires internet access as well as a mains supply.
- Chapter 6 gives details of how to set up a recorder, how to make best use of its features and has advice regarding video processing.
- Chapter 7 gives step-by-step instructions detailing how to set up various CCTV systems.
- Chapter 10 has suggestions on various species to monitor with CCTV.

Chapters 2, 5, 9 and 12 are more advanced and for those who wish a greater depth of information and more detail about techniques. They include scientific case studies intended for those involved in research.

- Chapter 2 can be treated as a CCTV reference and gives in-depth detail.
- Chapter 5 is on portable CCTV which is necessary for monitoring situations in the wild where trail cameras are not suitable.
- Chapter 9 describes two conservation case studies: a fish study conducted for Natural England and a pioneering study monitoring bats for the Woodland Trust.
- Chapter 12 is an advanced topic on triggered digital cameras when high-quality images are required.

1. Analogue CCTV

1.1 Analogue CCTV Quick Start

The commonest use of CCTV for wildlife monitoring is the camera bird box. Many get great pleasure from watching young birds in the nest, from blue tits to barn owls. For the novice, a bird box kit is the best way to start with CCTV. This section gives details of how to set up the sort of simple kit that is available from a wide range of suppliers.

1.1.1 Bird box – a simple introduction to CCTV

Figure 1.1 shows a basic CCTV set up.

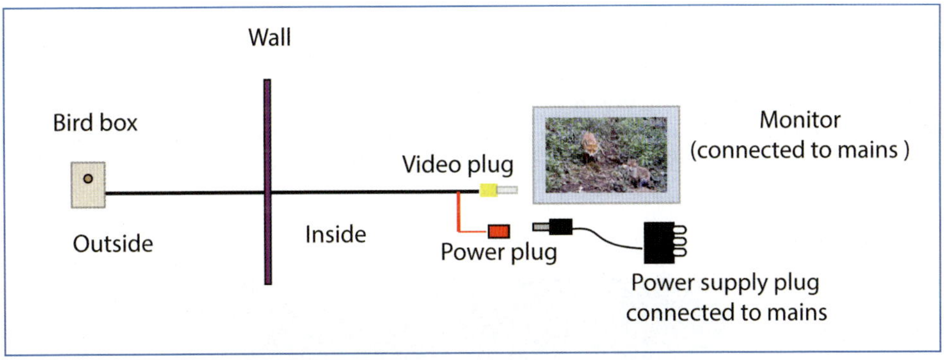

Figure 1.1 Basic analogue CCTV set up.

Step1. Buy a bird box kit (see Figure 1.2) consisting of:

- Camera (fitted in the box).
- Cables with RCA plugs.
- Bird box.
- Power supply adapter.
- SCART to RCA adapter.

Step 2. Fix the bird box containing the camera to a suitable tree (see Figure 1.3) or other surface.

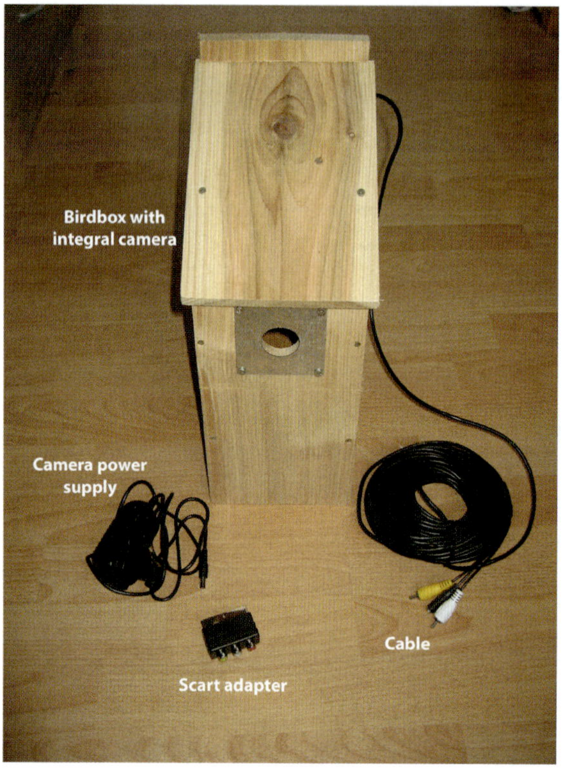

Figure 1.2 Bird box kit.

Step 3. Run the cable coming out of the bird box into your house. You may have to drill a hole in a wall or window frame. If you have power in a garage or shed, it is often possible to pass the wire past the door hinges or under the door. Cup hooks are good for keeping the wire out of the way (see Figure 1.4)

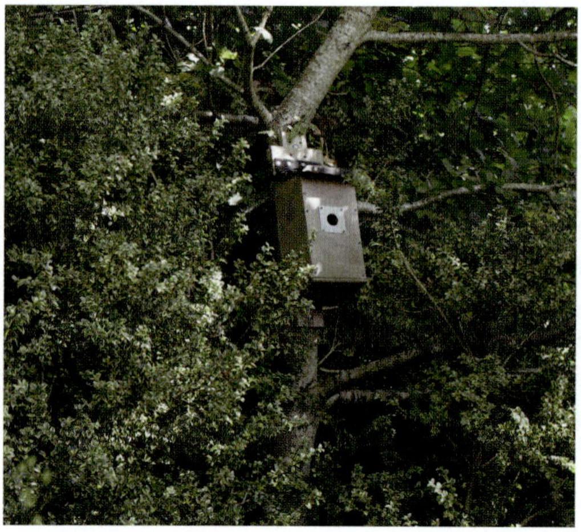

Figure 1.3 Bird box on tree.

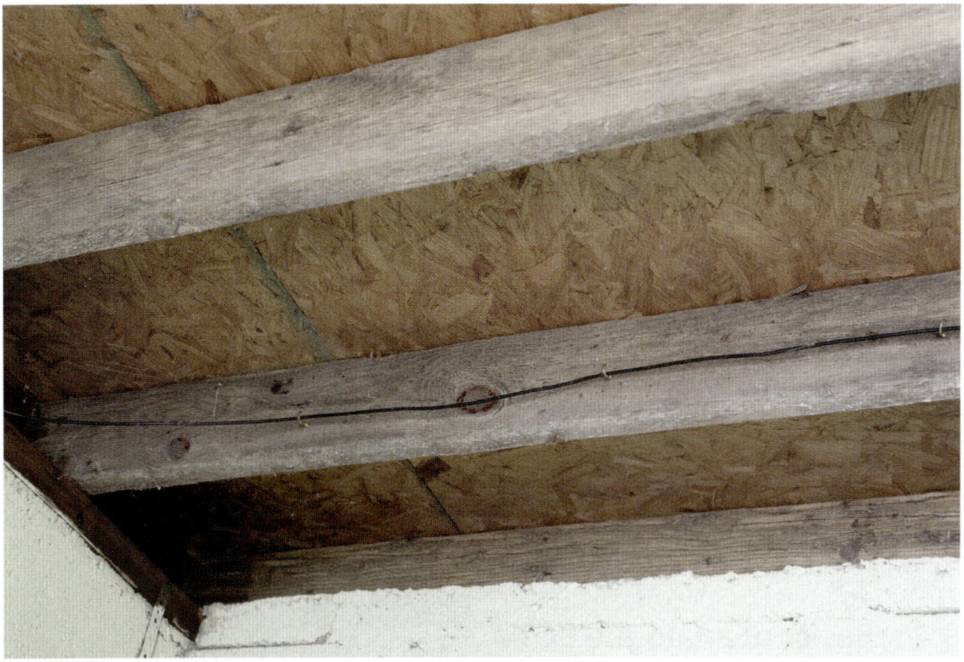

Figure 1.4 Cup hooks are good for organising the cable.

Step 4. Connect the camera power plug (red or black) to the power adapter, which should then be plugged into the mains.

Step 5. Connect the camera signal plug (yellow) to a monitor (also the white audio plug, if there is one). Some TVs have RCA sockets (yellow and white), but if your TV has a SCART socket you can connect to it using a SCART adapter. You can also buy monitors with RCA sockets.

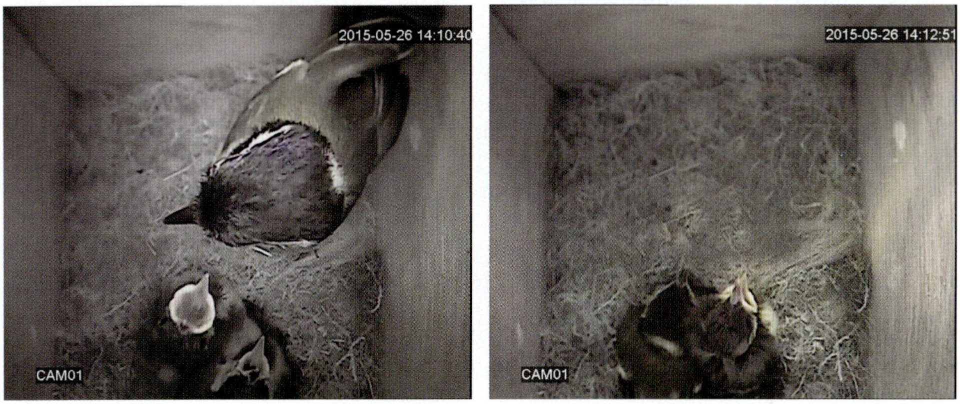

Figure 1.5 Bird box nestlings (video stills with a low-resolution camera).

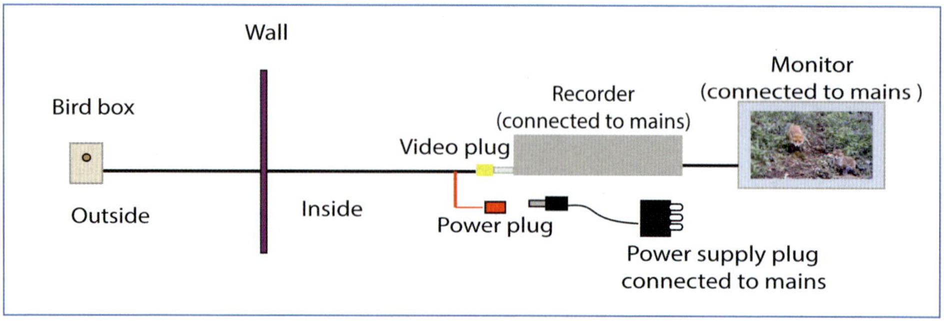

Figure 1.6 Basic analogue CCTV set up with recorder.

Whenever you switch on the camera power and the monitor, you will be able to watch what is happening inside the box (see Figure 1.5).

If you connect a recorder (see Figure 1.6) as well as a monitor, you can record videos. Get into the habit of selecting and exporting the videos daily.

1.2 Basics of analogue CCTV

1.2.1 Introduction

1.2.1.1 What are CCTV cameras?

CCTV is short for closed-circuit television. CCTV cameras differ from the cameras delivering television programmes in that the signal is confined to a small area. CCTV cameras are either directly connected with wires to a television, monitor or recording device, or transmit the video signal wirelessly over short distances (generally less than 100 m). They are widely used for security surveillance, but are also suitable for monitoring wildlife. Unlike trail cameras, they film continually. The sensors in the camera usually detect both visible and infrared light. At night, infrared light-emitting diodes (LEDs) usually supply illumination.

Most CCTV cameras are analogue, which means they transmit a continuously variable signal, unlike digital IP cameras (Chapter 4) which transmit a digital signal consisting of individual pulses.

Standard analogue CCTV cameras are best used for portable CCTV (Chapter 5) where there is no mains power present. For situations where mains power is present, HD analogue cameras will give a high-quality signal, with HD-TVI (Chapter 3) being the newest form of HD analogue CCTV. HD-TVI uses the best technology from analogue and IP cameras to give superb quality at an affordable price.

1.2.1.2 What do they look like?

For the purposes of this book, the CCTV cameras described are those I have used for wildlife monitoring, but there is a huge range of cameras available, due to the size of the surveillance market.

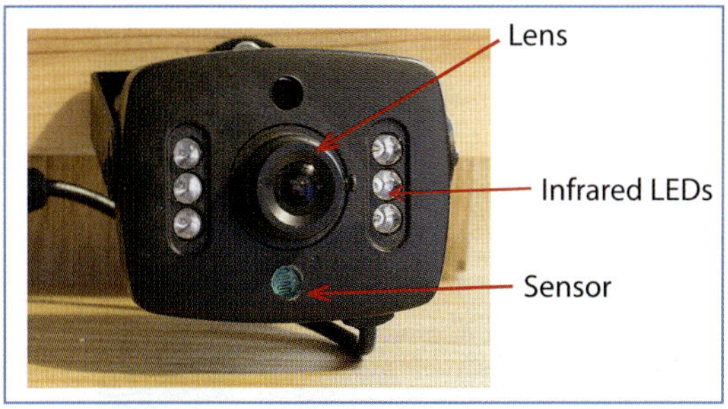

Figure 1.7 Bird box camera parts.

Typical small camera

Figure 1.7 shows a typical bird box camera, which can also be used for mammal boxes and in confined spaces. The sensor will detect when light levels fall and will switch on the LEDs.

Typical connections for a small analogue camera

The connections for a small analogue camera are shown in Figure 1.8:

- The red barrel connector carries power to the camera.
- The yellow RCA connector carries the video signal from the camera.
- The white RCA connector carries the audio signal (although not all cameras produce an audio signal).

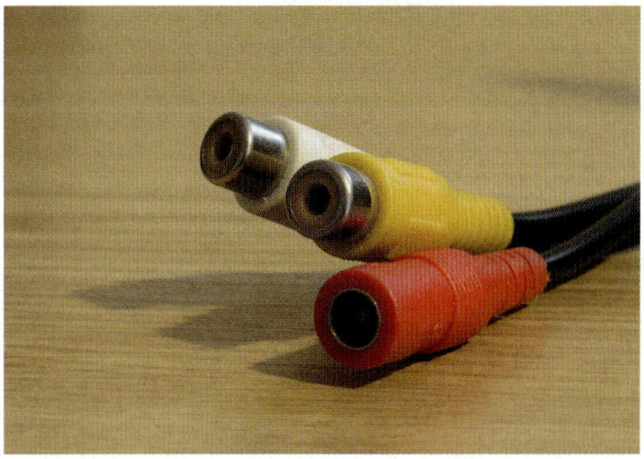

Figure 1.8 RCA camera plugs.

Typical bullet camera

Small 'bullet' cameras, useful for larger outdoor subjects, are available in a wide range of sizes and types. Figure 1.9 shows a typical CCTV bullet camera with the main parts labelled.

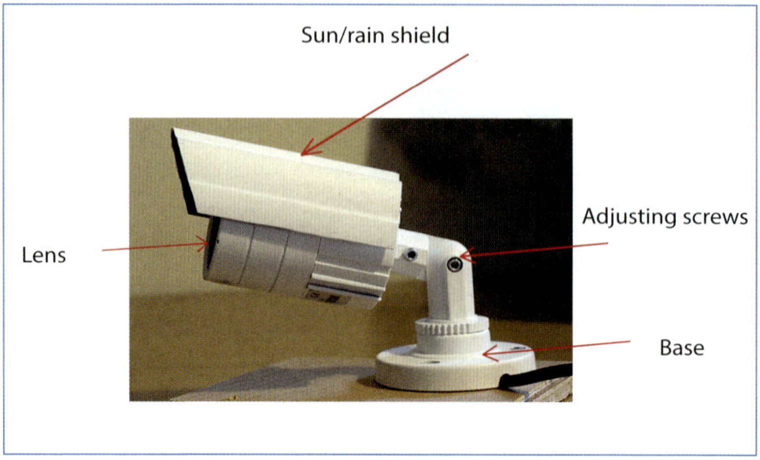

Figure 1.9 CCTV bullet camera parts.

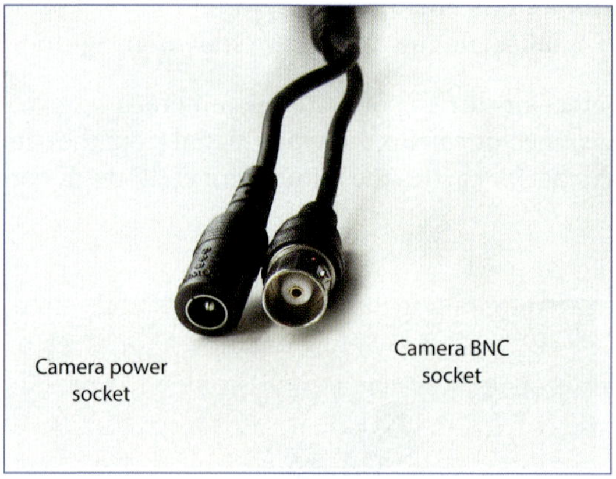

Figure 1.10 Typical bullet CCTV camera connectors.

Typical connections for a bullet CCTV camera

The connections for a small bullet camera are shown in Figure 1.10:

- The barrel power socket delivers power from a power supply to the camera.
- The BNC socket delivers the video signal from the camera.

1.2.1.3 Uses of CCTV cameras

Although the most popular use of CCTV cameras is to monitor activity in nest boxes, hedgehog boxes and bat boxes are also suitable subjects. The advantage of this use of CCTV is that once the cameras are set up, there is no disturbance to the wildlife. The main features of standard analogue CCTV can be summarised as:

- Cheap cameras available.
- Good for bird boxes and small mammal boxes.

- Best for portable CCTV.
- Night viewing possible with infrared illumination.
- Image quality average to good.

1.2.2 Basic analogue CCTV setup

The basic components of a simple analogue CCTV system are shown in Figure 1.11.

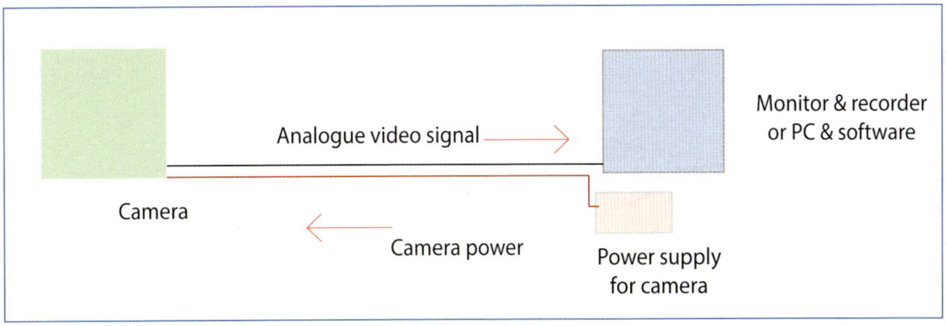

Figure 1.11 Analogue CCTV block diagram.

1.2.2.1 CCTV component details

The camera

CCTV cameras consist of a lens which focuses light onto a sensor. The electronics converts the signal from the sensor to an analogue video signal. Audio is sometimes available as well. The details are shown in Figure 1.12.

Wildlife CCTV cameras often need an infrared LED light source as many UK mammals are nocturnal. During the day, sunlight is reflected from the subject and the camera gives a colour image. If light levels are very low, the camera uses its own infrared light source and this gives a black and white image.

A CCTV camera needs power, usually in the form of 12 volts DC. Cameras kits supply a power adapter which can be plugged into the mains and which converts AC mains voltage to 12 volts DC for the camera.

There is a huge range of CCTV cameras to choose from. In practice, using a wildlife camera supplier is advised initially, especially as they have suitable connection kits and suitable outdoor cameras. A camera with the highest resolution (number of TVL lines) is advised.

Monitor

A monitor is required to view the images. This can be a TV or a separate monitor (such as used for a PC). The monitor is also required while setting up the camera, to make sure that the camera displays the area you wish to observe. The cable from the camera carrying the video (and in some cases audio) signal is plugged into the monitor.

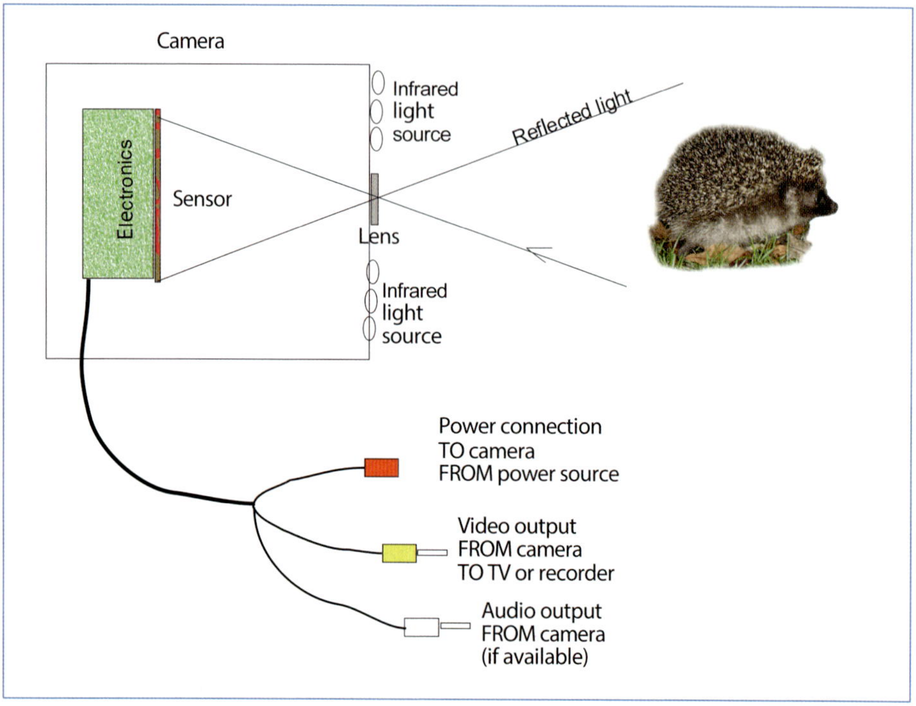

Figure 1.12 CCTV camera details.

Recorder

A recorder is required only if you wish a permanent record of what you see. The best recorders have motion detection and only record if the image from the camera shows movement. Note this is different from trail camera motion detection as that depends on changes in heat, rather than actual motion as here.

PC video capture

Instead of a recorder and monitor, a PC can be used. This requires a capture device to convert the analogue video signal to a digital form for the PC. Capture devices often come with software which allows viewing of the video directly on the PC. If more than one camera is viewed, this method becomes more complex, requiring a capture card to be connected inside the PC.

Software

One advantage of connecting cameras to the PC is that you can use software to control the recording. Motion detection recording is usually found on recorders, but using software on the PC gives much more flexibility in setting up the recording parameters. iCatcher is suitable software, and for one or two cameras is not expensive. It is not straightforward to set up, but support is available from the authors of the package (iCode).

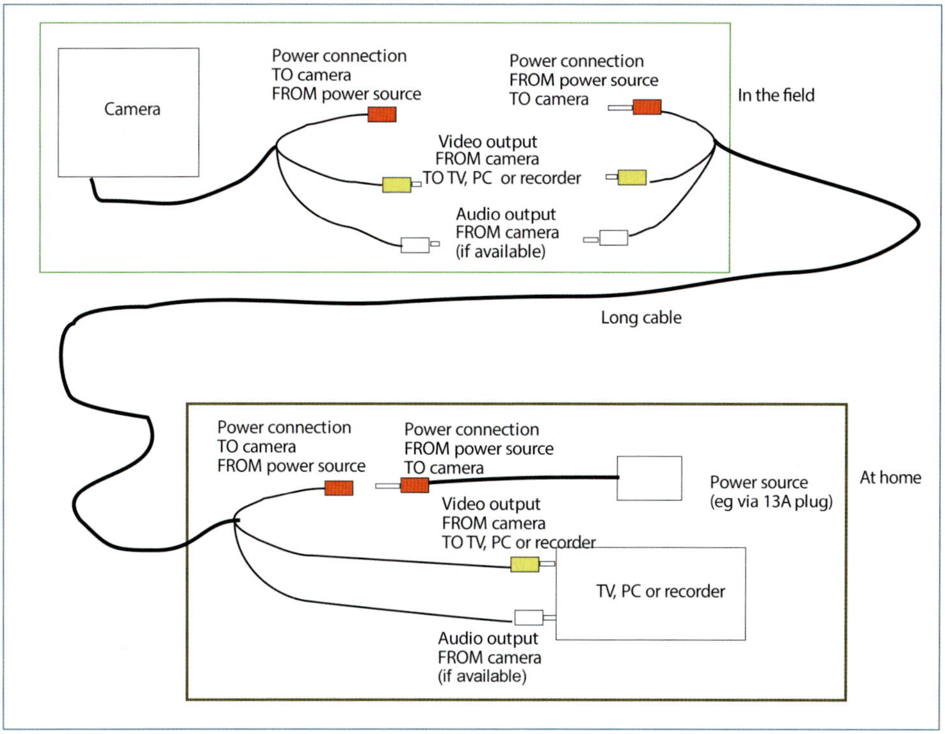

Figure 1.13 CCTV connections.

1.2.2.2 Connections

Details of how to connect a simple CCTV system from a kit are shown in Figure 1.13. Connecting up the system is straightforward with a kit. The connectors are colour coded and the connectors match in type or have suitable adapters. Most kits supply an audio connection (usually white) but this is not always needed as not all analogue CCTV cameras have sound.

A long cable connects the camera outside to the power source and recorder inside. Keeping the cable as short as possible is advised as this gives better video quality. Cameras can be connected via the cable directly to any TV or monitor that has the correct sockets, and most kits supply adapters.

Connecting to a recorder

Figure 1.14 shows a block diagram of a CCTV system with a recorder. If you use a recorder, the video cable is connected to the recorder input, often called 'AV in', and another cable is connected from the video 'AV out' to the monitor. The video files can usually be transferred to a USB drive and played on a PC. A recorder with a built-in monitor makes life easier but increases the cost.

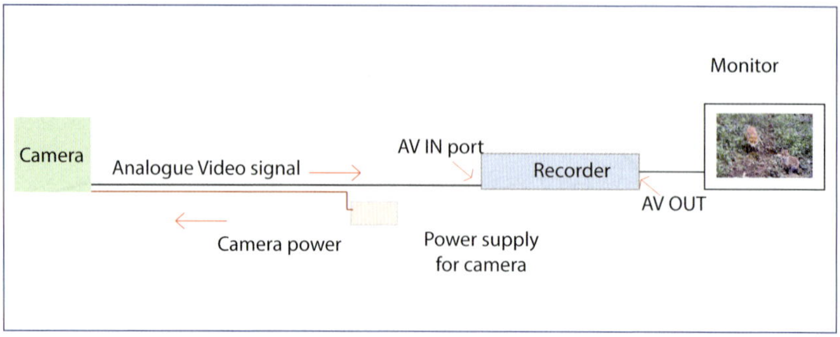

Figure 1.14 Analogue CCTV with recorder block diagram.

Connecting to a PC

A camera cannot be connected directly to a PC as the analogue signal has first to be converted to digital. Small video capture devices are easily obtained and these connect to the camera cable at one end, and the PC USB port at the other (see Figure 1.15). The devices come with a disc so that viewing software can be installed on the PC.

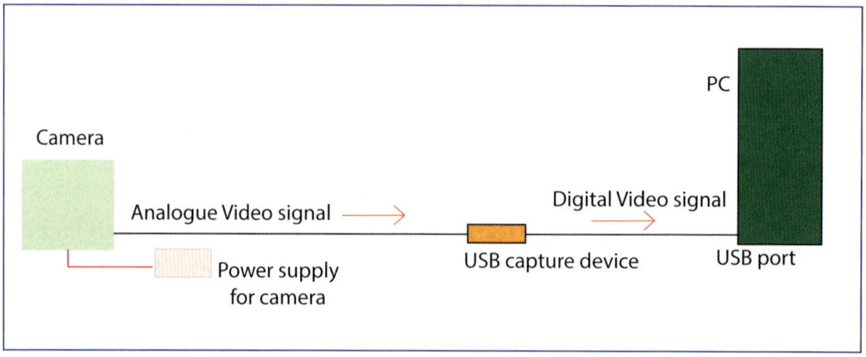

Figure 1.15 USB video capture.

1.2.2.3 Weatherproofing

It is important to find out if your chosen model of CCTV camera is waterproof, weather-resistant or not suitable for outdoor use. The more expensive type of CCTV camera used for general surveillance is usually weatherproof, but this is not always the case for nest box and other cameras, so care is required when buying. Cameras described as *weather-resistant* can usually withstand some rain, but for heavier bursts, a roof or three-sided box may be advisable (see Figure 1.16).

The better cameras use an IP rating where IP stands for International Protection marking. The most suitable rating for cameras are IP65 to IP68. The first digit, 6, means it is dust tight, the second digit describes how much water it can withstand. IP65 means

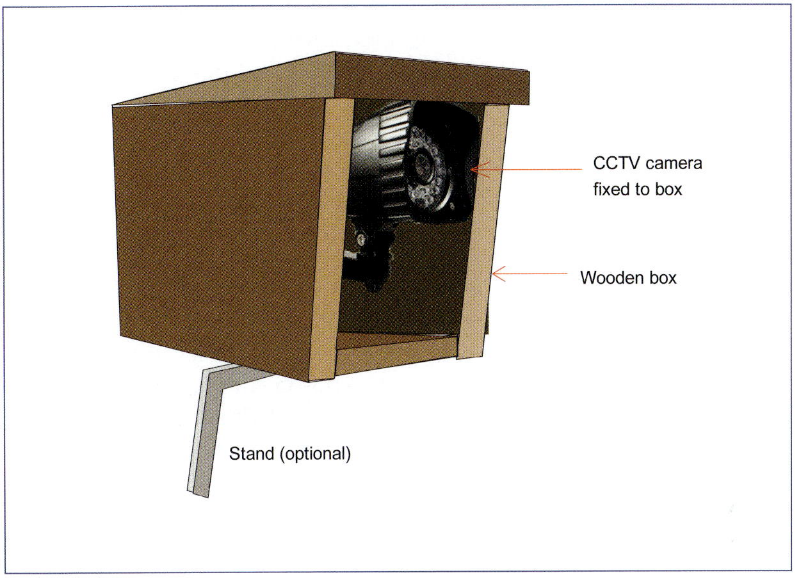

Figure 1.16 Camera protection box.

the camera can withstand jets of water and IP68 means the camera can be submerged. Further information from the camera manufacturer will be required as to how deep that submersion can be.

Thus IP65 to IP68 will be waterproof in heavy rain.

1.2.2.4 Power delivery

All CCTV cameras need a power source. Most CCTV cameras are connected to mains power (wired), but some are wireless and this affects the way in which power is delivered to the camera.

- *Wired cameras* In a wired system there is usually a wire for power attached to the video wire and this second wire has one end connected to the camera. This means that either the camera must be close to the location of the recorder or display device, or long cables will be required. The power is delivered using an adapter which is plugged into the mains and is connected to the other end of the power wire.
- *Wireless cameras* In a wireless system, although the signal from the camera is transmitted through the air, power is still required by the camera. This can be supplied via a cable, thus removing the advantage of wireless systems, or by a battery, which will need regular recharging.

1.2.2.5 Cost

The cost of a CCTV camera depends on its size and signal quality. Small nest box cameras can cost from £20 to £100, with the more expensive, high-resolution cameras costing £50 to £200 and upwards.

The image quality of a camera is described in term of TV lines (TVL). An explanation of this can be found in the section 2.5.1. Cameras with 600 TVL are standard analogue cameras and are not expensive. Those with 1,000 TVL are described as high-resolution and are a little more expensive.

1.2.2.6 Examples of use

Bird boxes (small cameras)

Bird boxes with built-in cameras are the easiest way to start with CCTV, but kits vary greatly in quality. For an easy start, make sure the camera is already fixed into the bird box and attached to the cable. That way, all you have to do is drill a hole to bring the cable from the outside to inside, then connect to a power supply, TV, monitor or recorder. Some boxes have visible lights as well as the infrared lights on the camera. The visible lights should have an in-built controller which switches off the power to the visible lights when it is dark, thus ensuring birds are not disturbed at night when resting.

Once you are familiar with the setup of nest box systems, it is not difficult to make your own bird box camera system. (See section 2.12.1 on making a camera bird box.) Be aware that most of the small cameras used in bird boxes and mammal boxes are not waterproof, so cannot be used outdoors without some protection.

Mammal boxes (small cameras)

It is possible to buy kits for hedgehog boxes and bat boxes, which have an almost identical setup to that for bird boxes, the only difference being the shape of the box and the positioning of the camera. It is not difficult to make your own mammal box and great fun can be had making boxes for mice, voles and shrews and watching their behaviour. Wooden boxes look good from the outside, but it is much easier to use plastic boxes, which have the additional benefit of being watertight.

The cameras with kits tend to give lower-quality images (see Figure 1.17). The quality of the image depends on several factors: camera type, camera quality, camera resolution, cable type, connections and lighting. Once the initial thrill of seeing 'your' birds or small mammals has passed, you may want to improve image quality. Ways to do this are described in section 7.1, and an example is shown in Figure 1.18.

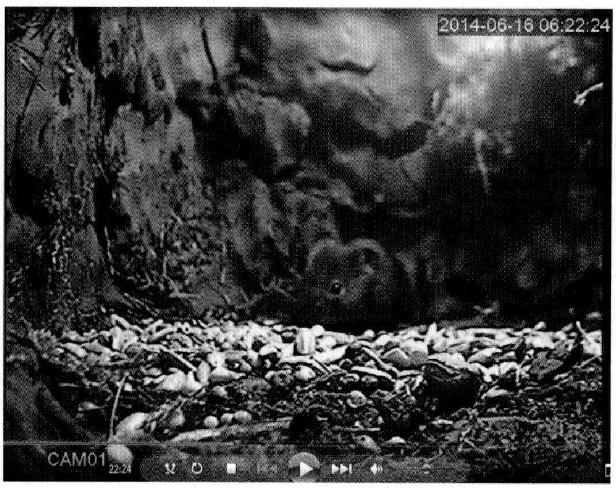

Figure 1.17 Video still of a vole using a low-resolution camera.

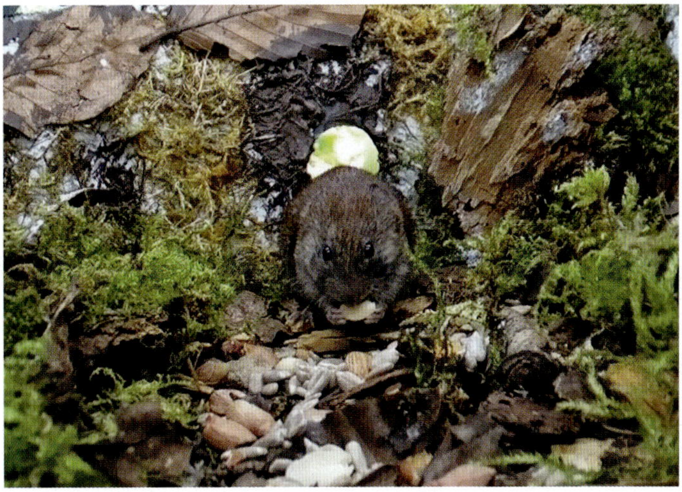

Figure 1.18 Video still of a bank vole (*Clethrionomys glareolus*) using an IP camera.

1.2.2.7 Modes of use

Colour videos by day (larger cameras)

Many UK mammals, such as deer, foxes, otters and badgers, are nocturnal, but, with the possible exception of badgers, most can be found out and about during the day if they think there are no humans (or dogs) around. Setting up a video camera in an area where mammals are likely to appear can be fruitful and can supply colour videos. However, unless you are lucky enough to have such areas on your doorstep, you may have to set up in a remote area, and trail cameras or portable CCTV systems will be required. You may have seen the result of CCTV cameras on television wildlife programmes, but you are

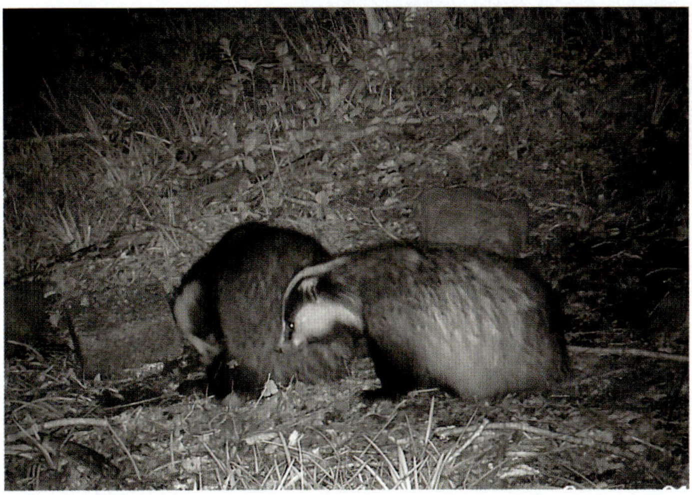

Figure 1.19 Video still of a sleepy badger mother and her cub taken using the integral LEDs of the camera.

not usually shown the kilometres of cables behind the scenes. A recent portable system behind the scenes of such a programme used a fuel cell 'battery' to supply power. Fuel cells can cost upwards of £5,000. The chapter on portable CCTV shows how a portable CCTV system can be achieved at relatively low cost.

Infrared videos by night

For those mammals that are nocturnal, most videos will be taken at night, with light supplied by the infrared LEDs on the camera. Additional light sources can be added if power is available. These can be white light from LED floodlights or infrared from illuminators. The clarity of the image is generally improved with additional light. However, good-quality CCTV cameras benefit from a good source of light provided by their integral infrared LEDs (see Figure 1.19).

1.3 Videos

Bird box

Figure 1.20 shows an image from a video of two nestlings in a camera bird box being fed. Because the camera is small, low resolution and does not have professional standard cables, the image quality is not as good as it could be, but the first video is always an achievement for the novice.

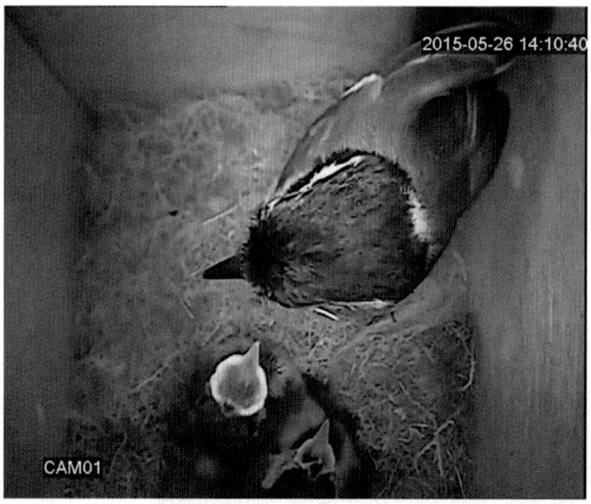

Figure 1.20 Bird box video thumbnail.
https://youtu.be/iXeLk76tR68

2. Advanced CCTV details

This chapter can be used as a reference. You may wish to skip it and come back to it when you want to delve deeper.

Once a basic system has been setup using a cheap kit, and you are familiar with its use, you may wish to improve the quality of your images. This chapter delves into CCTV more deeply and gives details on how to improve video quality.

2.1 Introduction

Cables are required to carry the video output from the camera to the monitor or recorder. If the camera has sound capability, a separate lead carries the audio signal. For basic CCTV systems, the video signal is carried by a basic cable with yellow RCA connections, and the sound cable connections are white, although not all cameras have a sound output.

For professional CCTV systems more advanced cables and connectors are used to give a higher quality of video, although audio is less common.

In most cases, CCTV also uses wires to transfer power to the camera. Although most cameras can be powered by battery, this can be inconvenient for continuous use, although necessary for portable CCTV.

Choice of cables and connectors has a major influence on video quality.

2.2 Digital and analogue

CCTV cameras today use similar sensors to those used in digital cameras and these sensors produce a continuously variable (i.e. analogue) signal. Digital still cameras convert the analogue signal from these sensors to a digital signal consisting of individual pulses, whereas, with the exception of IP cameras, CCTV cameras remain analogue.

If the signal from an analogue CCTV camera is transmitted to a recorder (as well as a monitor), the analogue signal is converted to a digital signal by the recorder and stored on digital media, usually an SD card or a hard disc.

2.3 Wireless CCTV cameras

The cables used for CCTV cameras are generally 20 m long or less, but they can be up to 100 m without appreciable signal loss. HD-TVI cameras can use up to 500 m of RG59 cable without loss. There are occasions when long cables can be a disadvantage and wireless camera systems may appear to be more convenient. In wireless systems, the video (and audio) signals are fed into a transmitter which transmits them to a receiver placed close

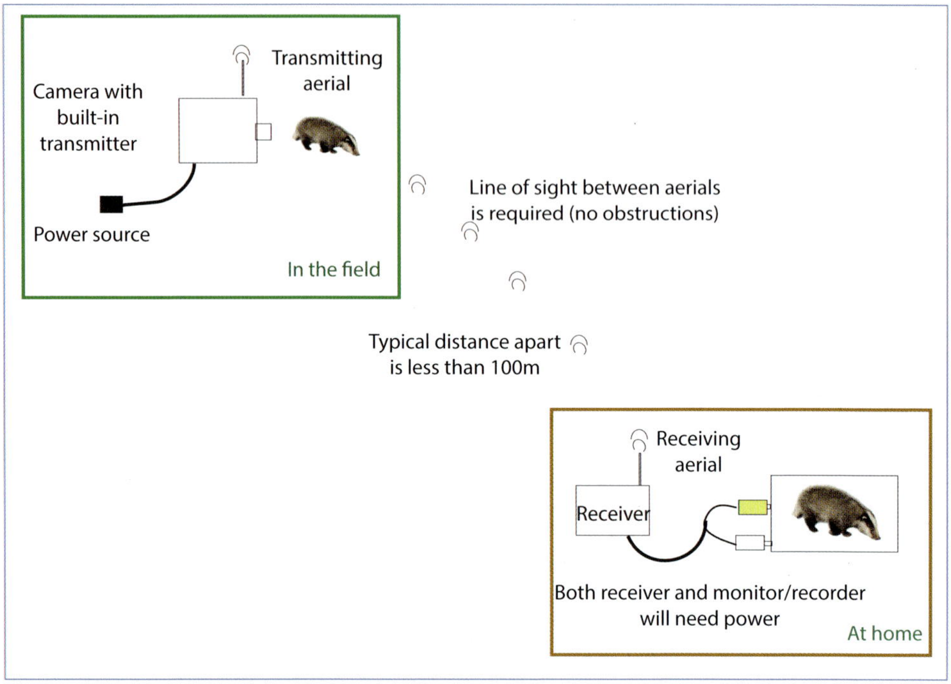

Figure 2.1 Wireless CCTV system.

by (such as in the house) without the need for cables. The receiver then transfers the signals to a monitor or recorder.

However, there are potential problems with wireless systems, both with supplying power and with interference. Power is still required for the camera and the transmitter, and unless mains power is available at the camera, batteries must be used. A power cable can be run to the camera, but that defeats the purpose of a wireless system.

Most wireless CCTV systems (see Figure 2.1) work on a frequency of 2.4 GHz, which is the same frequency as that used by broadband, microwaves and cordless phones. Because of this, care has to be taken with wireless technology as the signal can be hacked and other wireless signals, such as from a broadband internet system, can interfere with the signal.

In addition there must be a clear line of sight between transmitter and receiver for the signal to be transmitted without loss.

2.4 Power sources

2.4.1 Wired systems

For wired CCTV systems, the camera, recorder, monitor and PC use mains power. A source of power is required both in the field for the camera, and at home for the recorder, monitor or PC. If mains power is not available in the field, a portable system powered by batteries must be used. See Chapter 5 on portable CCTV.

2.4.1.1 Bringing power to you

If the area where you want to put your cameras is not close to your house, you might consider laying power cables to a shed or hide. The cost will depend on the distance, as the cables for mains power are best laid underground, so a digger must be used if you wish to avoid the problems of digging a long, deep trench by hand.

This is the option I chose as the woodland area where I wanted to place my cameras was too far away to use CCTV cables, and there was no clear line of sight for the wireless option. I also wanted to be able to plug in floodlights.

In my case a 120 m long trench was dug and power connected to a hide at a cost of £600. I opted for a power source of 4 kW, giving me the option for heating and additional lighting. For lower-power options, the costs are much less as thinner, cheaper cable can be used.

If you have access to a digger, and can lay the cable, you would only have to pay for an electrician to connect up the power, plus the cable cost. If you plan to have a number of cameras in use for a number of years, this may be a viable option. You can also keep a recorder in the shed and view images at your leisure. It is important, however, to ensure that any electrical equipment is protected from damp. This is not a problem in summer, but certainly may be in winter.

2.4.2 Wireless systems

For wireless systems, power is required for the camera and transmitter in the field. Batteries can be used for the camera and transmitter, if mains power is not available. Mains power is required in the home for the receiver, recorder, monitor or PC.

The type of battery that can be used to power a camera and transmitter will depend on the type of camera and transmitter used. Most will work with batteries, but some will not. Thus if a wireless system is to be used, care should be taken when choosing equipment.

2.4.2.1 Rechargeable C or D cell batteries

There are some cameras which have been adapted to use rechargeable C or D cell batteries. These systems are expensive, heavy and unlikely to last more than a few hours before needing recharged, which is not only inconvenient, but may disturb the wildlife being monitored.

2.4.2.2 Rechargeable lithium batteries

Rechargeable 12 V lithium batteries can be used for many cameras. These are light and convenient but can be expensive. The rating of a battery in ampere hours (Ah), gives an indication of how long the battery will last.

2.4.2.3 Rechargeable lead acid batteries

The safest form is a sealed type that will not spill acid. These batteries are heavy but are much cheaper than lithium batteries of the same capacity. They usually supply 12 V DC, which is the voltage that most CCTV cameras require, and are readily available. Lead

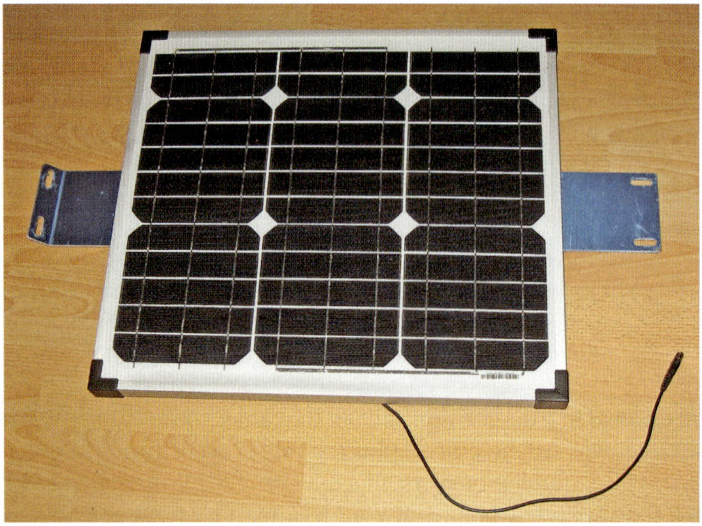

Figure 2.2 A 30 W solar panel with a mounting bracket attached.

acid batteries described as suitable for electronic components are more likely to supply the smooth voltage required for the CCTV system.

Solar panels

Solar panels (see Figure 2.2) are useful for prolonging battery charge and there are accessories available that will allow the use of solar panels to charge the batteries used by the camera. They can be mounted on a wall or pole with a suitable bracket. However, solar panels cannot be used in shaded areas, which can limit their application.

2.5 Image quality

The image quality of CCTV cameras depends on many factors: camera quality, number of lines (TVL), sensor used in the camera, light levels, focusing of lens, quality of recorder, length of cables, whether wired or wireless, whether or not there is electrical interference. Section 7.1 gives details of the equipment and techniques required to obtain the best quality possible.

2.5.1 Camera quality

It is difficult to judge camera quality without seeing the output, but generally speaking, cheap cameras will not give as good a signal as more expensive cameras, although that is not always the case. In order to choose, you should study the supplier's website carefully and judge whether its claims are justified. A small nest box camera can be purchased for £20; a high-spec waterproof CCTV camera can cost more than £200. Added to that is the cost of cables, power source and various connectors. As stated in Chapter 1, for an initial purchase, buying a complete kit with a nest box is probably the most economic method and has the advantage of making clear exactly what components are required. There are several good companies to be found on the internet. After the first purchase, buying the best you can afford may be the way to go.

2.5.1.1 Resolution: PAL

In order to judge the importance of the number of lines quoted with a camera's specification, it is important to understand exactly how the analogue video signal appears on a monitor. The picture information is 'painted' on the monitor screen as the analogue electrical signal moves across and down the screen (see Figure 2.3). The UK uses PAL technology which means that the signal moves 625 times across and down the screen.

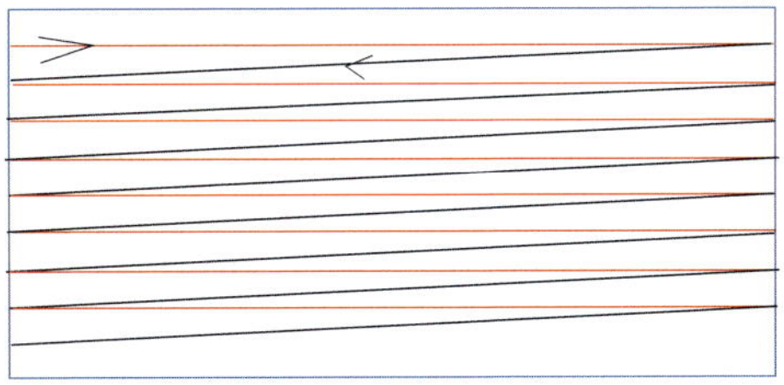

Figure 2.3 The signal moves across the screen (red lines) then quickly back (black lines). The process repeats until 625 horizontal lines have been traversed on the screen.

The PAL system thus describes the vertical resolution as 625 lines. Although the lines are horizontal, the term 'vertical resolution' is used as the lines are stacked vertically down the screen.

2.5.1.2 Resolution: number of lines

Video camera quality is quoted in terms of TVL or television lines, with 1,000 lines being high resolution and 300 being low resolution. This describes how well the camera (or monitor) can allow the viewer to distinguish two lines that are close together as shown in Figure 2.4.

In Figure 2.4, when using the same monitor, all cameras should be able to show two separate lines A, a mid-range camera should be able to show two separate lines B but only high-resolution cameras will be able to show two separate lines C. A resolution of 600 TVL means 600 vertical lines can be seen separately (300 dark and 300 white). With a resolution of 1,000 TVL, it is 1,000 lines.

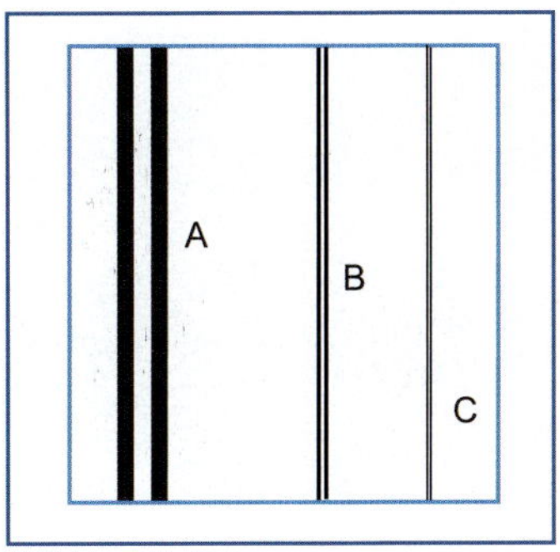

Figure 2.4 Resolution is the ability to separate two lines.

All CCTV cameras have sensors, and resolution is determined by the number of pixels on the sensor, as with digital still cameras, so the term TVL can be confusing to those used to digital cameras. The term TVL is still used because the cameras transmit an analogue signal. 600 TVL is equivalent to a 0.5 megapixel sensor, and 1,000 TVL approximately to 1.4 megapixels. In general, the higher the TVL, the better the image.

2.5.1.3 Light levels

As with any camera, light levels have a major effect on CCTV image quality. For cameras in the open during daylight hours this is less of an issue, but for cameras in nest boxes, light levels are something to consider. Nest boxes can be made with transparent plastic windows, allowing light to enter the nest box, and it is also possible to add some LED lights which are powered by the camera supply. However, care should be taken that there is not so much light that the birds are disturbed or they may simply leave. For this reason, side windows alone are advised and are sufficient for daytime unless the nest box is in a very dark corner.

For the majority of outdoor monitoring applications, bullet cameras with integral LEDs should be used. Cameras usually state the range of the LEDs used, and this should be considered when choosing a camera. A range of 20 or 30 m is suitable for many wildlife applications, as animals filmed at a greater range would be rather small.

2.6 Connections and wiring

Connections and wiring can have a major effect on video quality, and once you have bought your camera, it is important to learn as much about these as possible, if image quality is important to you.

Chapter 1 described the equipment available for wildlife CCTV, but these typical systems will not generally give good-quality images, even with good cameras. The reason is that the connections and cables supplied are often cheap and easy to assemble, but are not generally those used by professional CCTV systems for surveillance. Without delving into the theory of image transmission in depth, choice of cable and connectors is key to improving image quality.

Most good-quality cameras transmit video with a screened coaxial cable and (usually) BNC connectors, all with 75 ohm impedance, impedance being equivalent to the electrical resistance of the circuit. If the camera is attached to a recorder using RCA connectors and

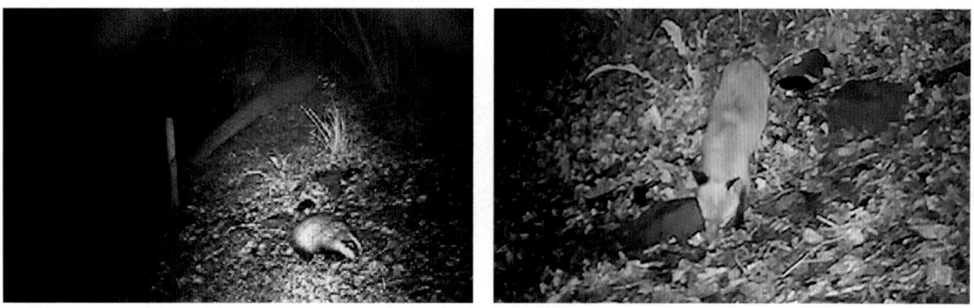

Figure 2.5 Video from (left) basic CCTV wiring and connectors and (right) well-matched connectors.

cheaper cable, the image data is likely to be compromised, which means that some of the information may be lost or affected by electrical noise and/or interference.

The images in Figure 2.5 show the difference between a well-matched cable/connector system (on the right) and a typical 'amateur' system (on the left).

2.6.1 Cable

2.6.1.1 RG59 coaxial cable

The top wire in Figure 2.6 carries the video signal and would usually be connected to a male BNC plug. The red and black wires in the smaller cable carry power and are usually connected to a barrel male plug or female socket.

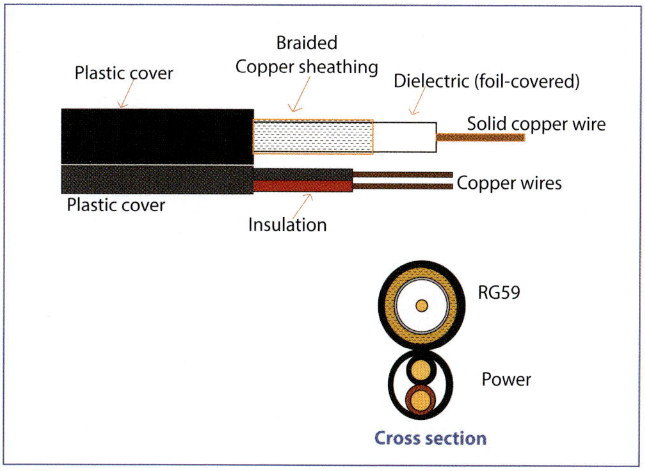

Figure 2.6 RG59 cable with power cable attached (Siamese cable).

For the best images, good-quality coaxial cable such as RG59, which has an impedance of 75 ohm and is screened to reduce interference, should be used for image data transmission, together with good-quality BNC connections also with a 75 ohm impedance. Matching impedance, as this is called, is important as it reduces signal losses.

Even RG59 cable comes in different qualities and low-loss cable using copper wires will give a better result than standard cable. If you have cable runs of more than 30 m, low-loss RG59 cable should be used. Cables should be as short as possible, and with as few connections as possible.

2.6.1.2 Cat5 cable

It is also possible to use Category 5 (Cat5) cable, which is a 'twisted-pair' system, rather than RG59.

Cat5 cable consists of four pairs of wires in the colours shown in Figure 2.7. Copper conducting wires are insulated in solid and striped insulation of each colour (blue, green, brown and orange), and are twisted together. Noise from interference in one cable is cancelled out by the noise in the paired cable.

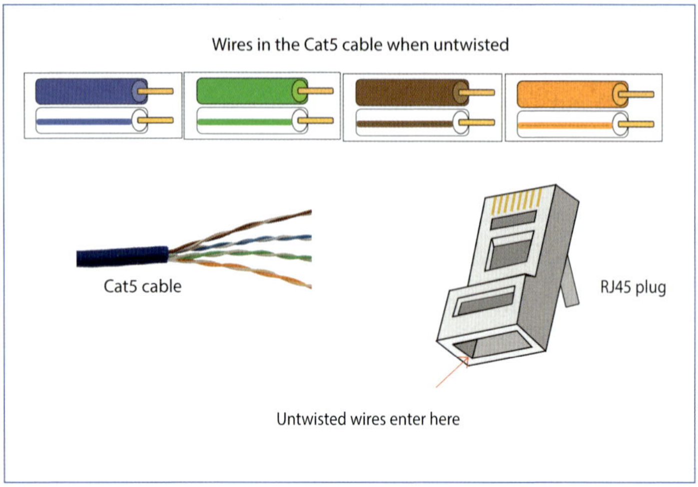

Figure 2.7 Cat5 cable with an ethernet RJ45 plug.

Cat6 cable is a better-quality cable with increased screening and waterproofing.

The complete assembly of Cat5 cable with an RJ45 connector at each end is also called an ethernet cable. Ethernet cables are used for network connections, and for example, carry broadband signals from a router to a computer.

For this reason, Cat5 cable is used with IP CCTV cameras (Chapter 5), but it can also be used with analogue CCTV.

2.6.1.3 *Cat5 cable with analogue CCTV*

A balun is defined as a simple transformer and in CCTV is a more specialised form of adapter. CCTV baluns (see Figure 2.8) are useful for adapting the BNC connection from

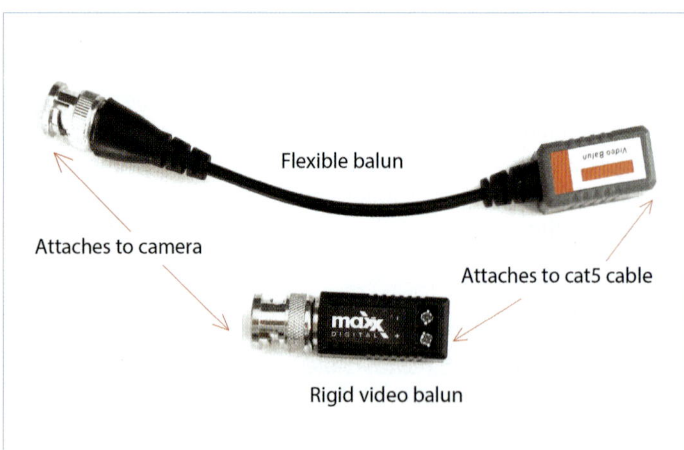

Figure 2.8 CCTV video baluns. Either flexible or rigid baluns can be used to adapt a BNC connection so that it can be used with a Cat5 cable.

a CCTV camera so that it connects with a Cat5 cable and will also match impedance. The male BNC on the balun connects to the CCTV camera signal lead, and one of the Cat5 wire pairs is connected to the other end of the balun simply by using a screwdriver. This is repeated at the opposite end of the Cat5 cable with a second balun.

It does not matter which wire pair is used as long as there is consistency at each end of the cable. For example, the solid orange should be connected to the positive connection (+ sign) on the balun at each end of the cable, and the striped orange to the negative (– sign) at each end.

The advantage of using Cat5 cable instead of RG59 cable is that only a screwdriver is required to make up a cable.

2.6.1.4 *Power baluns*

A power balun is a very simple adapter (see Figure 2.9). Most power cables have connectors moulded to either end, but sometimes there is a need to attach a power barrel connector to bare wires. If wires from the Cat5 cable are to be used to carry power alongside the wires carrying the video signal, a power balun will be required together with the video balun at each end of the Cat5 cable. The bare wires are pushed into holes at the open side of the balun and fixed in place with screws.

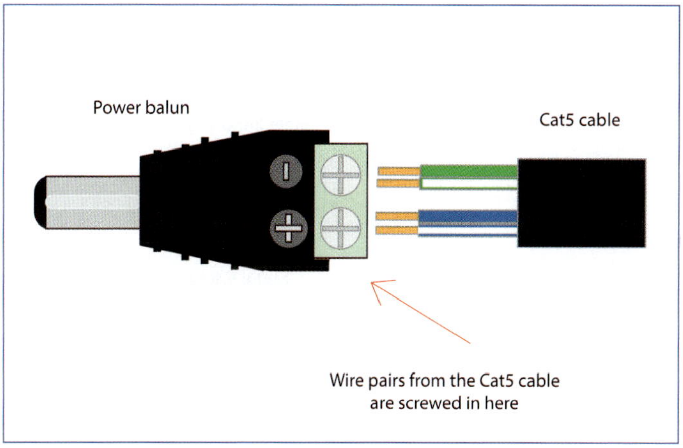

Figure 2.9 Power balun with a male plug used with a Cat5 cable.

Two wire pairs from the Cat5 cable can be used for power, with one pair connected to the positive connection on a power balun, and the other pair to the negative. The third pair is used for the video signal and the fourth pair can be cut off.

Power baluns come in various forms, with Figure 2.9 showing the simplest. It has a 2.1 mm barrel plug at one end, and the Cat5 cable power wires are screwed in the other end.

A power balun with consistent connections is required at each end of the cable. This means that if the blue/blue-white-striped pair is connected to the positive screw at one end, it must also be connected to the positive screw at the other end of the cable. Normally the other end of the cable uses a female power balun, which has a female socket, instead of a male plug.

2.6.2 Connectors

There is a wide variety of connectors which vary according to the camera type and the monitor. Each connector has a male and female part, with the male part having a pin and the female a hole into which the pin connects. They appear much more complex than they actually are, and simply transfer the image information from the camera to the monitor or recorder. There may be a separate connector carrying sound (if the camera is capable of recording sound) and a connector supplying power to the camera.

2.6.2.1 Phono or RCA connectors

RCA connectors (see Figure 2.10 and Figure 2.11) are often used in wildlife CCTV kits and are connected to basic cables. They are also used by recorders and monitors. These connectors are traditionally are colour coded for video use:

- Yellow for composite video signal (the camera data). This is an output connector as a signal comes *from* the camera.
- White for sound (if available from the camera). This is an output connector as a signal comes *from* the camera.

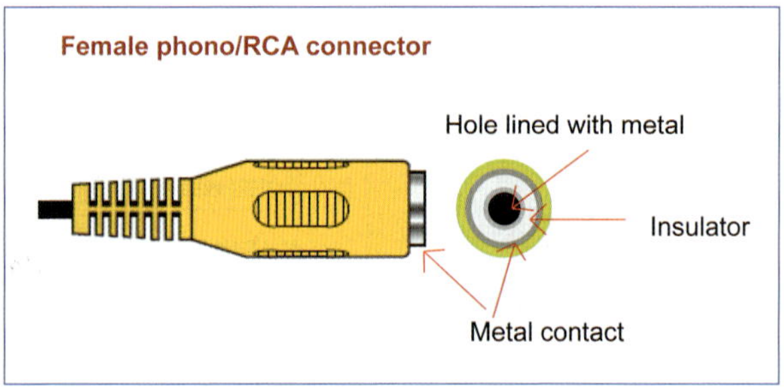

Figure 2.10 Female phono/RCA connector.

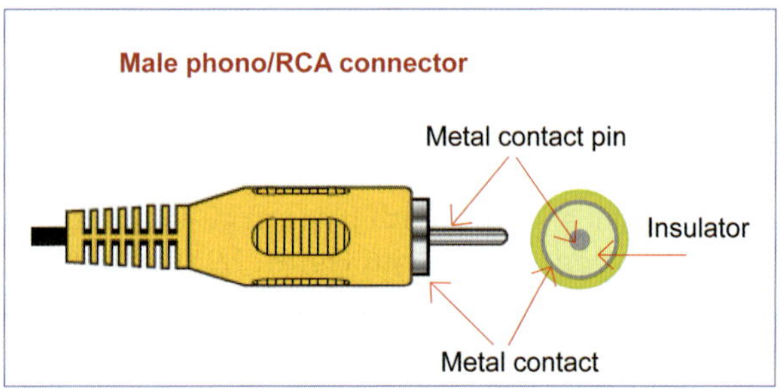

Figure 2.11 Male phono/RCA connector. The centre pin carries the video signal.

2.6.2.2 BNC connections

Higher-quality CCTV cameras are used in professional systems with a screened RG59 cable. For high-quality images, BNC connectors (see Figure 2.12 and Figure 2.13) should be used for the video signal.

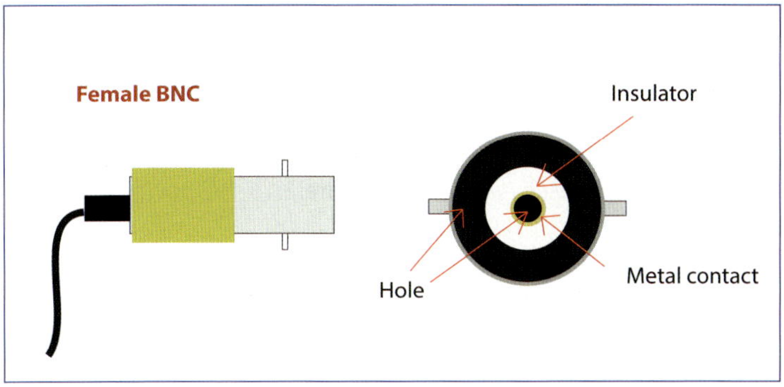

Figure 2.12 Female BNC connector.

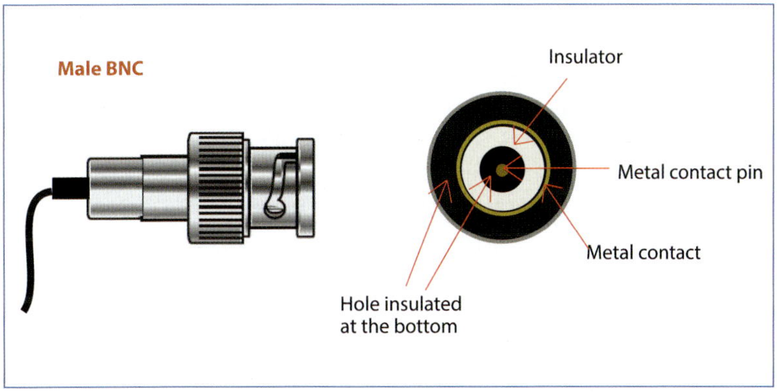

Figure 2.13 Male BNC connector.

2.6.3 Adapters

2.6.3.1 Power adapter

If you are faced with trying to connect two female (or two male) connectors together, an adapter connector may be required (see Figure 2.14). Beware: there are different diameters of jack power plug, so be sure to get the correct one. A typical size is 2.1 mm inside diameter, 5.5 mm outside diameter.

A situation that often occurs is that two devices, such as a camera and a portable recorder, have to be powered by a single battery. In this case a power splitter cable is used.

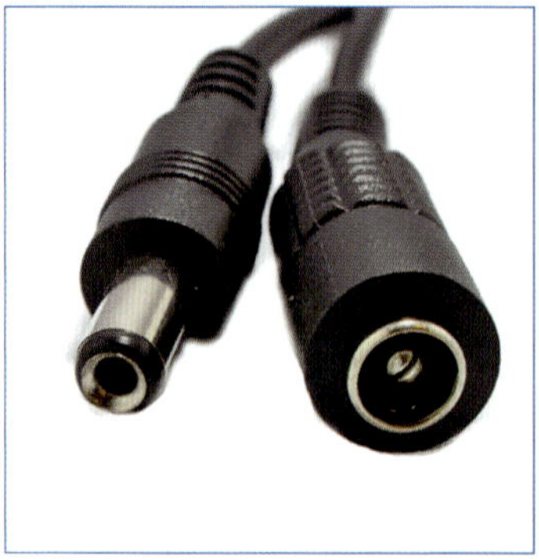

Figure 2.14 Power connector. Male jack on left, female socket on right.

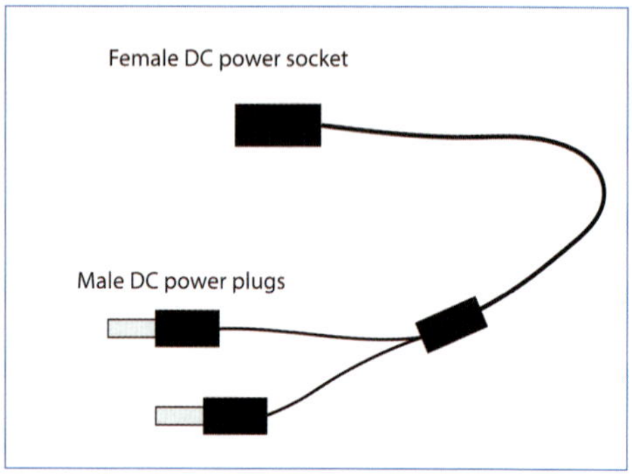

Figure 2.15 DC power splitter.

For example a lithium battery with a male power connector may be required to supply power to a DVR and a CCTV camera, each with a female power socket. In Figure 2.15 the female socket of the splitter would connect to the battery, and the male plugs would connect to the DVR and the camera. This way only one battery is required to power two separate pieces of equipment.

It is essential that both camera and recorder have the same voltage requirements, for example 12 V DC. The current drawn by each device must also be known so that the total current is less than the rating of the cable.

For example if the cable is rated at 1 amp, the DVR current plus the camera current must be less than 1 amp. Thus if the DVR current draw is 150 mA and the camera draws 350 mA, the total current drawn is 500 mA and the cable is suitable.

If the DVR current was 150 mA and the camera current was 900 mA, the total current would be 1,050 mA and a 2 amp rated cable should be used.

2.6.3.2 BNC and RCA adapters

Problems arise when connectors from the camera are different from those required by the monitor, TV or recorder. The commonest problem is trying to connect a BNC to a RCA connector, but adapters are commonly provided. Two examples are shown in Figure 2.16 and Figure 2.17.

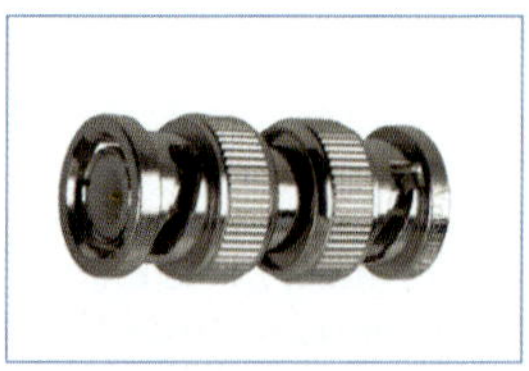

Figure 2.16 Male BNC to male BNC adapter, joins two female BNC sockets.

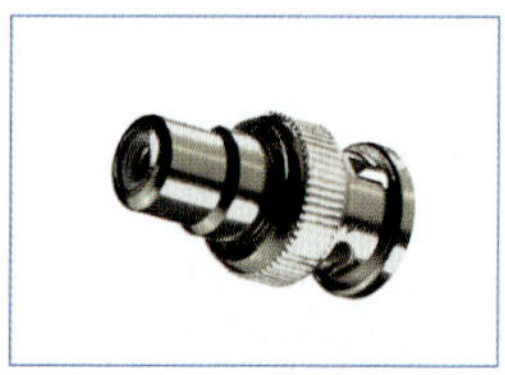

Figure 2.17 RCA female to BNC male adapter, joins a male RCA to a female BNC.

2.6.3.3 Camera to monitor connectors

Monitors have a range of input connectors. Sometimes they have RCA or SCART connections, but often they may be VGA or HDMI. In this latter case, a convertor may be required. These can be purchased cheaply online, but make sure you buy a convertor rather than an adapter. A convertor has electronics to convert the signal, whereas an adapter does not, and is unlikely to work.

2.6.3.4 Camera to recorder

Large recorders usually have female BNC sockets for video. The cable from the camera to the recorder should have male BNC connectors, which makes connecting straightforward, otherwise an adapter may be needed. Small, portable recorders may have RCA connections for which you will need a BNC to RCA adapter.

2.6.3.5 Camera to PC

As described in Chapter 1, if a connection to a PC is required, a capture device must be placed between the PC and the cable from the camera, so that the analogue signal is converted to digital. As a result, the camera-to-PC connection is actually a camera-to-capture device connection. Capture devices often have RCA sockets, so an adapter may be required. If you have several cameras, the situation is more complex, and a capture *card* is required. This is plugged inside the PC.

In order to avoid the complexities of cables and connectors, it is advisable to buy ready-made cables which will have (or should have) all you need, and hopefully find a knowledgeable salesperson to guide you. However, you may wish to make your own cables, which can save money once you know which parts to buy and are willing to learn how to make good connections. It takes a great deal of practice.

Unfortunately, very few of those who sell CCTV equipment for wildlife mention the importance of good connections, so you may be left wondering why your images are not as good as those the camera advertises. One solution is to opt for IP systems (see Chapter 4 on IP systems) as they are supplied with good-quality ethernet cables with very few connections. However, IP is only possible with power *and* an internet connection close by.

2.7 Camera parameters

2.7.1 Range

2.7.1.1 Nest box cameras

Range varies a great deal with cameras, depending on their use, and can be defined in several ways. The cameras used for nest boxes are designed to give a sharp focus a few centimetres from the bottom of a nest box, and may require a focal length of between 3 and 11 mm, depending on the box dimensions. In a nest box the required range is the distance from the camera to the bottom of the bird box.

With variable-focus cameras, the lens can be turned to move it in and out a small distance, allowing for some adjustment of focus to give a sharp image of the nest. This feature is useful if different sizes of box are used.

Depth of field is also a consideration. Depth of field is that distance over which an image is sharp once the lens has been focused on a certain point. For example, if the lens is fixed to give a sharp image 16 cm away, a camera may give a reasonably sharp image between 11 and 21 cm. In this case the depth of field of the camera lens is 10 cm.

2.7.1.2 *External bullet cameras*

For larger bullet cameras, focal length varies between 3 and 50 mm, with 11 mm being a typical value. In this case, the useful range can be considered as the greatest distance at which the subject can be seen clearly and will depend on the size of the subject. Thus the same camera may have a range of 3 m for birds and 20 m for deer.

If the focal length is increased, the image is magnified and the range will increase, so that, for example, birds 5 m away may be seen clearly in daylight.

Some cameras have a variable-focus lens, with a specific range over which it can be focused. CCTV cameras do not generally have automatic focus, and zoom and focus are adjusted manually. The more expensive high-end cameras may have remote focusing. This gives much more flexibility as a moving animal can be kept in focus for a greater distance, although the controls will require the presence of an operator.

At night, range is determined by the distance illuminated by the camera's infrared LEDs, and is around 30 m on average. However, the size of the subject to be filmed must be taken into consideration. A small subject may be illuminated at 30 m but would be too far away to see clearly.

As the price of a camera and the current it draws increase as the night range increases, care should be taken to use a camera with a night range suitable for the subjects to be studied. Cameras with a greater night range have more powerful LEDs. Taking a 'one size fits all' approach is not advised as, for example, a subject close by may be overexposed by a camera with a range of 30 m, whereas a larger, more distant subject may not be illuminated sufficiently to give a clear image.

2.7.2 Field of view

This term describes how much the camera sensor can see of its surroundings.

A camera with a wide field of view will see everything between the green lines in Figure 2.18 so that the trees with birds, pond and duck will be in the frame. A camera with a narrow field of view (the area between the blue lines) will see less, and only the pond and duck will be in the frame.

2.7.2.1 *The effect of focal length*

Field of view depends on the size of the sensor to some extent, but the lens focal length has the greater effect. Small nest box cameras are likely to have a shorter focal length lens because they have to focus on nest box contents only a few centimetres away, which means they have a wider field of view.

Deciding what you are likely to use the larger cameras for will help decide the field of view most useful. For example, if you are monitoring a fox den with a camera, you may only need a field of view of a few metres. If you want to monitor birds in your garden, you may want a wider field of view, but for watching feeders you may only want a field of view of less than 1 m. Thus when choosing a camera, the focal length and hence the

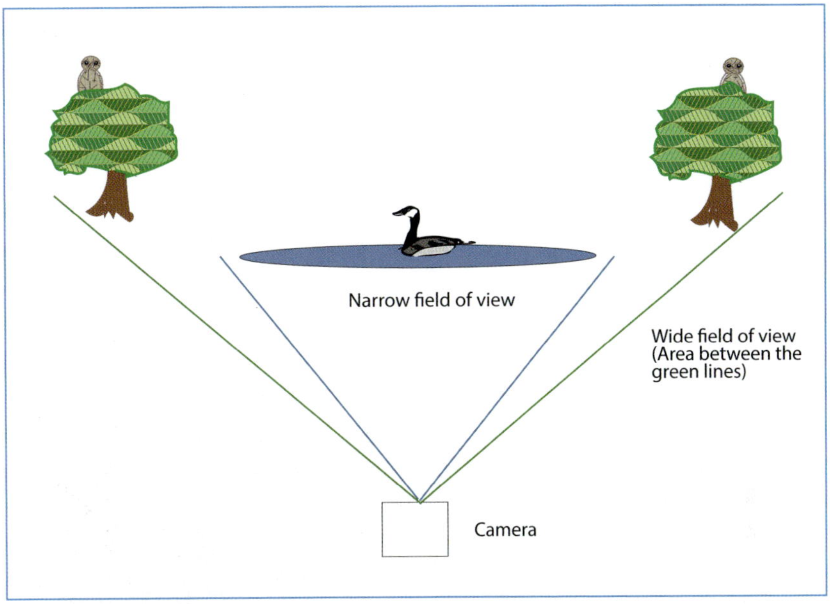

Figure 2.18 The difference between a wide and a narrow field of view.

field of view should be appropriate to the subject. For this reason, a variable-focal-length camera is more useful for monitoring wildlife than a fixed-focus camera. Field of view will change as the focal length is adjusted.

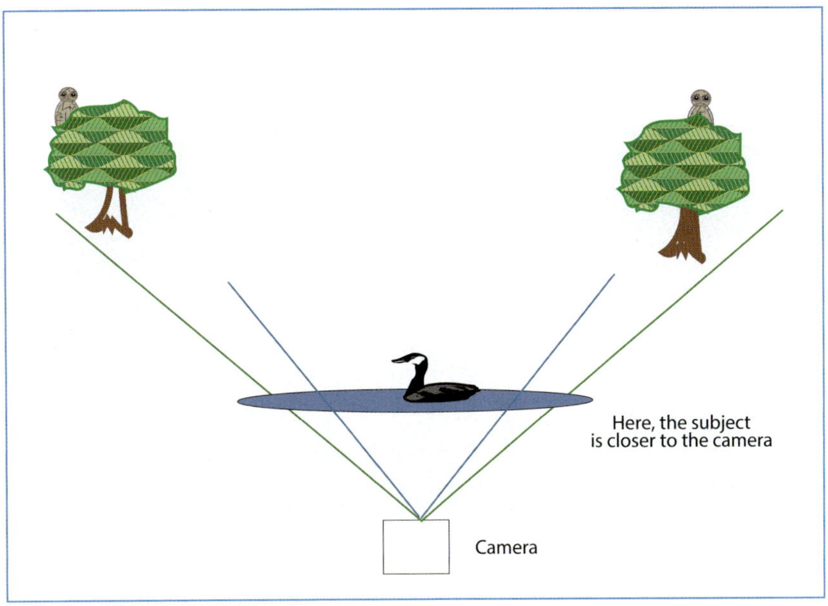

Figure 2.19 If the object is closer, the field of view seems narrower.

2.7.2.2 The effect of distance

If a subject is close to the camera, less will be seen than if it were further away from the camera. In Figure 2.19, even the camera with a wide field of view would see less of the pond if it was much closer.

2.7.2.3 Field of view examples

In Figure 2.20, the badger is a very small part of the frame. This was taken with a 2.8 mm lens.

When the lens focal length is adjusted to 11 mm (see Figure 2.21), the badger takes up more of the image.

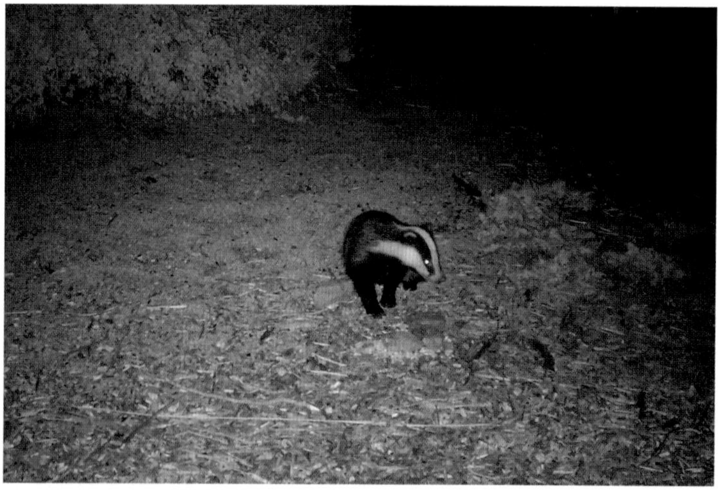

Figure 2.20 Wide field of view.

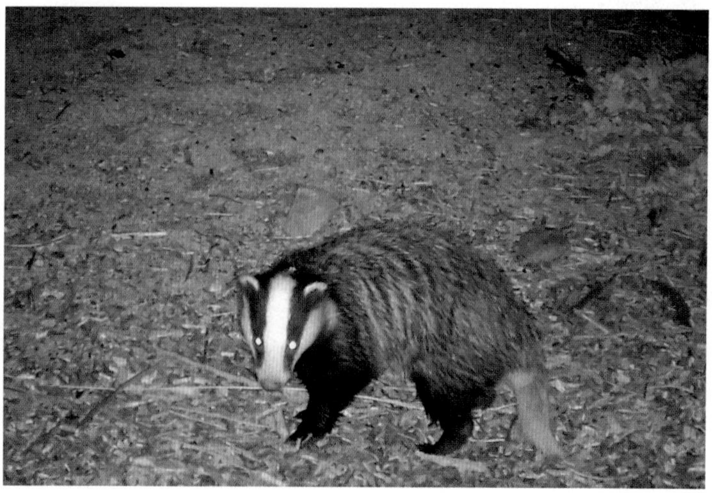

Figure 2.21 Narrow field of view.

2.7.3 Other parameters

2.7.3.1 Camera trigger speed

Because CCTV cameras are filming all the time they are switched on, time to switch on and start filming (trigger speed) is not relevant as far as the camera is concerned. Triggering is a feature of the recorder.

2.7.3.2 Cut filters

Most CCTV cameras for wildlife produce video only. Most have infrared illumination which switches on automatically when it becomes dark. The better cameras automatically place an infrared cut filter in position during the day and remove it at night. These 'true day/night' camera give better colour during the day

Cheaper cameras work without a filter, and daylight colour is distorted by the infrared element found in sunlight.

2.7.3.3 Audio

Wildlife CCTV cameras can be useful with sound and it is possible to obtain cameras that record audio as well as picture. If professional cables such as RG59 cables are used, a separate wire may be required to carry the audio signal.

2.8 Recorder parameters

The choice of recorder is very important, and for filming wildlife, there are several parameters to consider.

2.8.1 Motion detection

As CCTV cameras film continuously when switched on, there is no need for motion detectors to switch on the camera, as with trail cameras. The disadvantage of this is that any recording may have large periods with nothing interesting to show. This can be solved to some extent by the use of a recorder equipped with a motion detector, so that it records only when triggered by motion. This is different from the method used in trail cameras, where the camera switches on when triggered by heat changes. With CCTV cameras, the camera is always running and it is the motion detect feature of the recorder that triggers recording.

2.8.2 Pre-record

Some recorders have a pre-record feature when set up for motion detection. The recorder keeps 10 seconds or so in memory at any time, and if motion detection is triggered, it adds that 10 seconds before the recording. As a result, however fast moving a wildlife subject is, the recorder will have recorded its movement. Without this feature, any delay in recording when triggered could lead to the loss of a vital few seconds. This is a common problem with trail cameras, but is avoided with the pre-record feature of recorders.

2.8.3 Sensitivity of motion detection

The fact that actual movement triggers a recording when the recorder is setup for motion detection can be a disadvantage, as moving leaves or branches can start the recorder, giving rise to a large number of 'empty' recordings. Motion is determined by comparing groups of pixels, and if these change, recording starts. Reducing the sensitivity means that it requires larger groups of pixels to trigger the recorder which can reduce the instance of 'empty' videos. It is best to avoid placing the camera near moving vegetation when possible.

2.8.4 Masking

Many recorders have a masking feature. This allows parts of the view to be masked so that they do not trigger motion recording. Constantly moving items such as water and foliage can be masked out but the whole scene will still be seen on any videos. This can greatly reduce the instance of 'empty' videos.

However, even with low sensitivity and the use of masking, triggering may take place by minor events that cannot be masked, such as heavy rain, falling leaves and moving patches of light.

The movement of reflected sunlight in still pools of water can also be a problem. Careful siting of the camera and a reduction in motion detection sensitivity can help reduce 'empty' videos. If masking is not available, recording at night rather than round the clock may solve the problems caused by moving light patches.

2.8.5 Scheduling

If you wish to record continually without motion detection, the ability to schedule times during which recording will take place aids flexibility. Some recorders have the facility to record by motion detection within a scheduled time span. This is an excellent feature and affords greater control over recording.

2.8.6 Recording after the event timing

Some recorders make it possible to limit the length of time that recording takes place after triggering. It is much easier to analyse several short clips than one very long one, so this can be a useful feature. It is particularly helpful if the recording is triggered by a short event, such as a falling leaf, as it reduces the length of unwanted content.

A moving subject, such as a mammal, will keep triggering recording as long as it is moving, so clip lengths will be determined by the duration of the motion. For example, a recorder with a pre-record time of 10 seconds may be set up to record for 6 seconds after an event. A falling leaf video may only last about 16 seconds: pre-record time of 10 seconds plus post event recording of 6 seconds. A badger moving around for 75 seconds, will record a video of about 85 seconds: pre-record time of 10 seconds plus 75 seconds, the time for which it was moving.

2.8.7 Clip lengths

Afterwards, the recording has to be transferred to a PC. This is more convenient with a

USB flash drive or similar, provided that it has sufficient capacity for what can be large files. It is important to make sure that the video files produced are in a standard format so that viewing is straightforward.

Clip length should not to be confused with recording after the event timing. Most recorders limit the maximum clip length, and will split up a long event into several separate clips. This makes analysis easier.

2.9 Setting up

2.9.1 Positioning the camera

2.9.1.1 Mounting

Positioning the CCTV camera well is important. Cameras inside nest boxes are fixed to the box with screws (Figure 2.26 gives an example of how to mount the camera), but for free-standing cameras, some kind of support is required so that the camera can be fixed to that and then mounted where suitable.

Mounting methods for external cameras also depend on whether or not the camera is waterproof. There is not much point in using a camera outside if it is not waterproof. A camera may be described as waterproof with an IP rating of 65, but IP68 allows submersion, and is thus the highest waterproof rating.

You also have to bear in mind that connections between cameras and cables are not waterproof and they have to be well protected, either by using a junction box or by wrapping them with adhesive waterproof tape. Any water ingress to the connections is likely to cause interference or signal loss.

Mounting the camera on the side of a wooden shed or other building with the connections placed inside the building is the best method, if that is feasible.

2.9.1.2 Sun

Positioning the camera with regard to the position of the sun is not as critical as with trail cameras as there is no heat-dependent motion detector to trigger the camera. However, it is still important to ensure that the sun does not shine directly into the camera as that could damage the sensor at worst, and at best would overexpose the image.

2.9.1.3 Aiming the camera

As a motion detector is not used to trigger the CCTV camera as a rule, the angle at which the camera should be placed will depend on the surroundings. Pointing the camera up a known trail will show more action than if the camera is at right angles to the trail. If the target comes from behind the camera, then the target will be seen going away. If the trail is used on a regular basis, it is likely the first few sessions will make it clear whether to point the camera up or down the trail.

With the camera pointing up the trail as in Figure 2.22, but to one side (so that the target does not knock it over), the target will be seen for much longer.

If the camera points across the trail as in Figure 2.23, the target will only be seen for a short time when it passes into the field of view of the camera (between the purple lines).

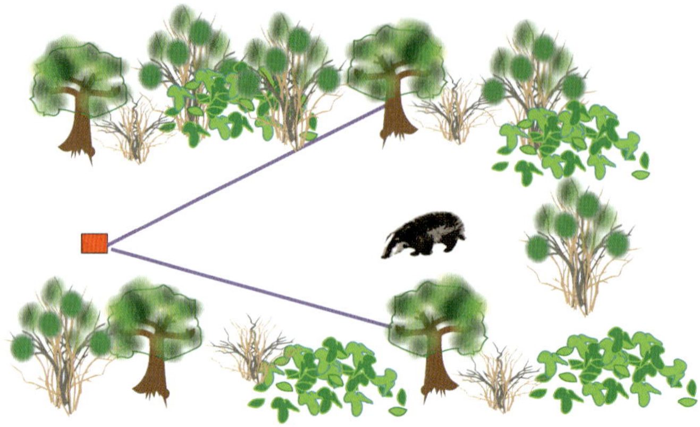

Figure 2.22 Camera pointing up the trail.

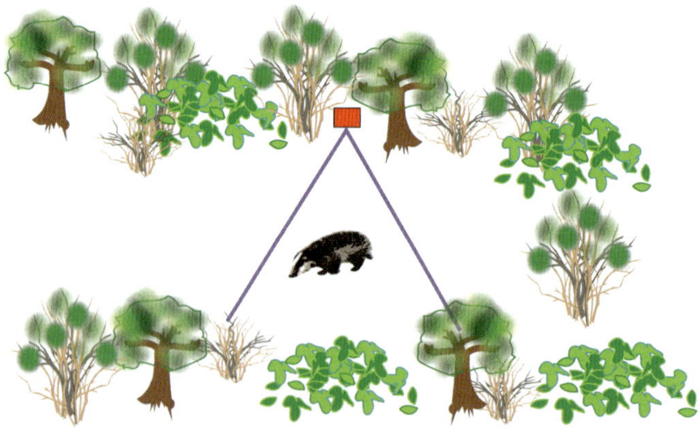

Figure 2.23 Camera pointing across the trail

2.9.1.4 Obstructions

If there is no known trail, the camera should be pointed in the direction with fewest obstructions. In Figure 2.24 the camera's view is obstructed by trees so the target will only be seen in the small triangular area.

Although the camera is pointed at right angles to the target in Figure 2.25, it is positioned pointing towards fewer obstructions, so the possible viewing area is larger than in Figure 2.24.

2.9.1.5 Vegetation

A CCTV camera should be placed so that vegetation does not obstruct its view as shown in Figure 2.25. In addition, if the recorder is in motion-detect mode, moving branches may start it recording and should therefore be avoided.

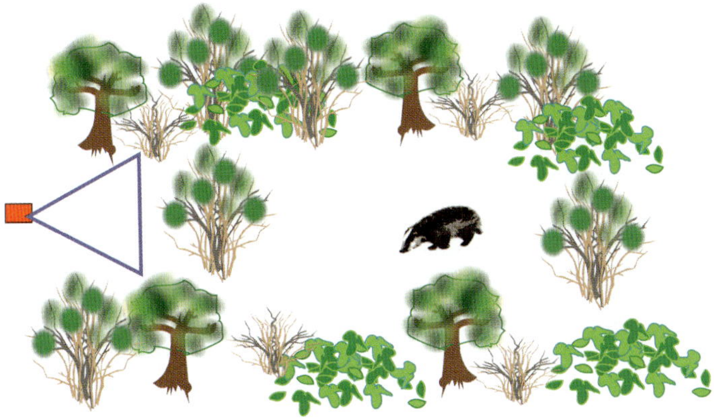

Figure 2.24 Camera view obscured by obstructions.

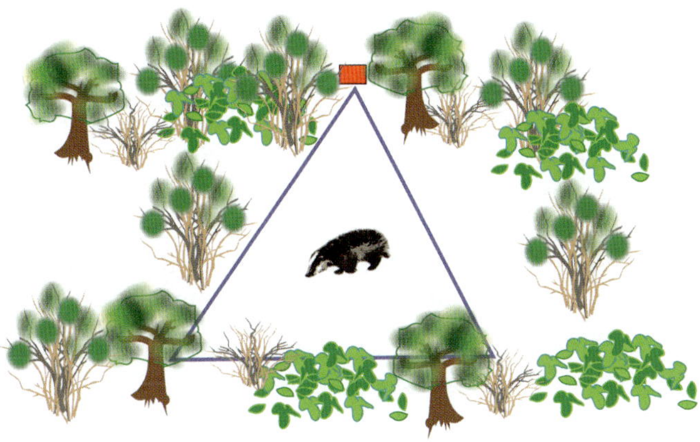

Figure 2.25 Fewer obstructions means a greater viewing area.

2.9.1.6 Height

The height at which the camera is mounted will depend on the target. For small mammals it can be placed low to the ground, and for larger animals, higher up. The flexibility of small cameras, such as those used in nest and mammal boxes, gives much greater scope for filming small animals and birds. CCTV cameras come in a wide range of sizes and are very flexible with regard to mounting position.

2.9.1.7 Angle

Most bullet cameras used for outdoor viewing can be moved up and down, usually by loosening a screw with an Allen key. Some cameras also have a ring at the base that can be loosened to allow the camera to rotate. If mounted on a horizontal surface, this means that the camera can move from side to side or pan, which allows precise adjustment.

However, if the camera is mounted vertically, on a tree for example, this motion only straightens up the camera.

The more expensive cameras sometime have two articulated joints, each of which can be loosened with an Allen key. This allows the camera to move up and down and from side to side while mounted vertically. This is a very useful feature and is worth paying for, as it makes precise camera positioning very much easier.

2.9.1.8 Avoiding theft

Theft is always a possibility in areas of public use. As CCTV cameras are usually screwed to a surface they are more difficult to remove. It is also possible that potential thieves may believe they are being filmed and may be less likely to attempt to steal the camera.

2.10 Fixing details

2.10.1 Large cameras

Many cameras have a base with screw holes, which can be screwed to a wall or tree. Others have a tripod-size screw socket and can be connected to a tripod or to a small CCTV stand. Small CCTV stands that are like small tripods are supplied with some cameras, but I have found that they are not always strong enough to support the weight of the camera without becoming loose.

A bracket is useful as it allows movement of the camera in several planes, which makes it possible to aim it exactly where you want. Some of the more expensive models can be moved by remote control, but these tend to be used for security rather than wildlife monitoring. A moving camera is likely to alarm wildlife, especially if its motion is not silent.

2.10.2 Small cameras

This category includes cameras used in nest boxes, and other small enclosed areas. Mammal boxes often have the camera on the side of the box, with the entrance hole directly opposite the camera. Nest boxes for birds are generally fixed to a tree, and the camera inside fixed to a shelf. The cameras in bird boxes can also be fixed to the side, but will have to be contained in another box for waterproofing. One method of attaching a camera is shown in Figure 2.26.

2.11 Other considerations

2.11.1 Using a monitor for testing

A monitor is always needed when setting up the camera. You have to be sure that the camera is focused so that the image is sharp, and that the camera is pointing in the right direction with a suitable field of view. In a bird box this means focusing a few centimetres from the base of the box. An artificial nest with artificial birds can be useful to get the correct focal length. Make sure you buy a camera that has the ability to adjust zoom and focus. This is often done by loosening small screws and turning them until the correct part

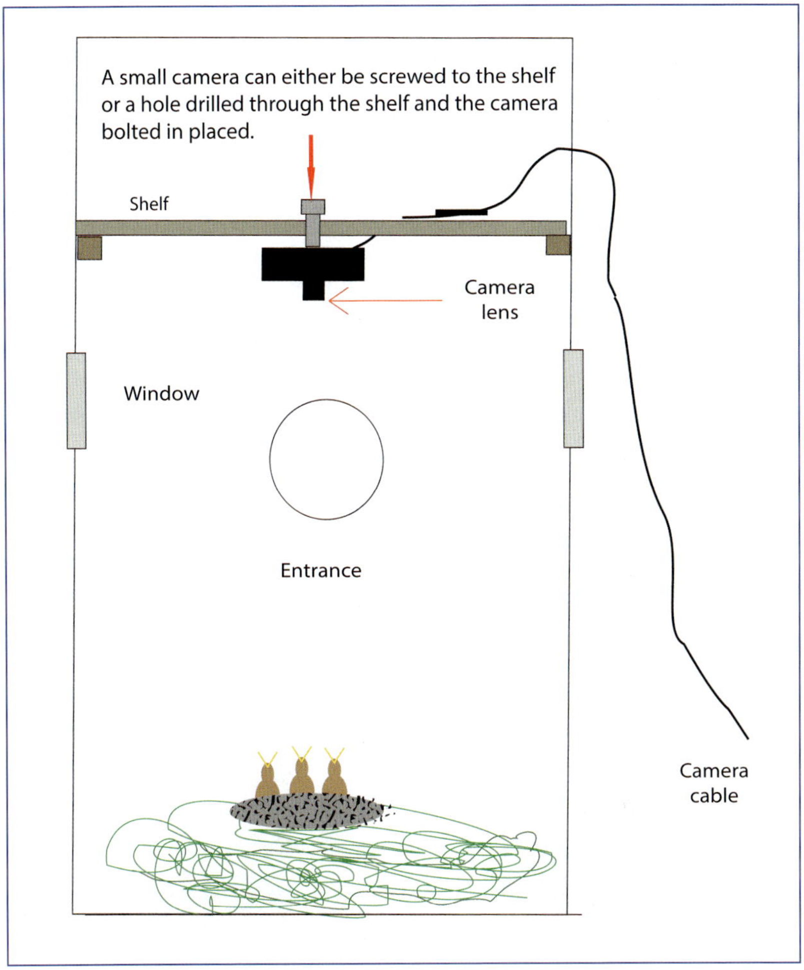

Figure 2.26 Fitting a camera inside a bird box.

of the 'nest' is in focus. The screws should be tightened and the artificial nest removed. Focusing can only be achieved accurately by connecting the camera to a monitor and studying the image.

For larger cameras outside, a portable test monitor is a very useful tool for ensuring the camera is positioned correctly. These small hand-held monitors have integral batteries and are indispensable.

2.11.2 Internet viewing

If you use a recorder with an internet interface, it can be connected to your internet router. Such DVRs use an app (often free) which allows you to view the DVR images on a smartphone, tablet or PC. If you wish to stream live to the internet in this way, it is advisable to do this via a streaming company to ensure you do not overload your internet traffic.

2.11.3 Maintenance

Wildlife cameras, especially in nest boxes, have a harder life than the digital camera you may have at home, so may not last as long. Maintenance is easier if the camera is attached to a wildlife box by enclosing it in its own box and fixing that to the wildlife box, be it bird box or mammal box. That way it is easy to remove and clean it, or to remove it when not in use.

The larger CCTV cameras are more expensive and built for longer life, but will be at the mercy of the elements. Often maintenance is more an issue of prevention rather than cure and it is vital to ensure that cameras are fully protected from the elements, either by buying a fully waterproof camera or by enclosing it in a weatherproof box. Cable connections should always be protected from rain and enclosing them in a plastic box can be useful, although adhesive insulation tape works well if checked regularly and renewed if it seems to becoming loose. Ensure that lenses are cleaned regularly with a soft clean cloth, especially as spiders often attach webs to the camera.

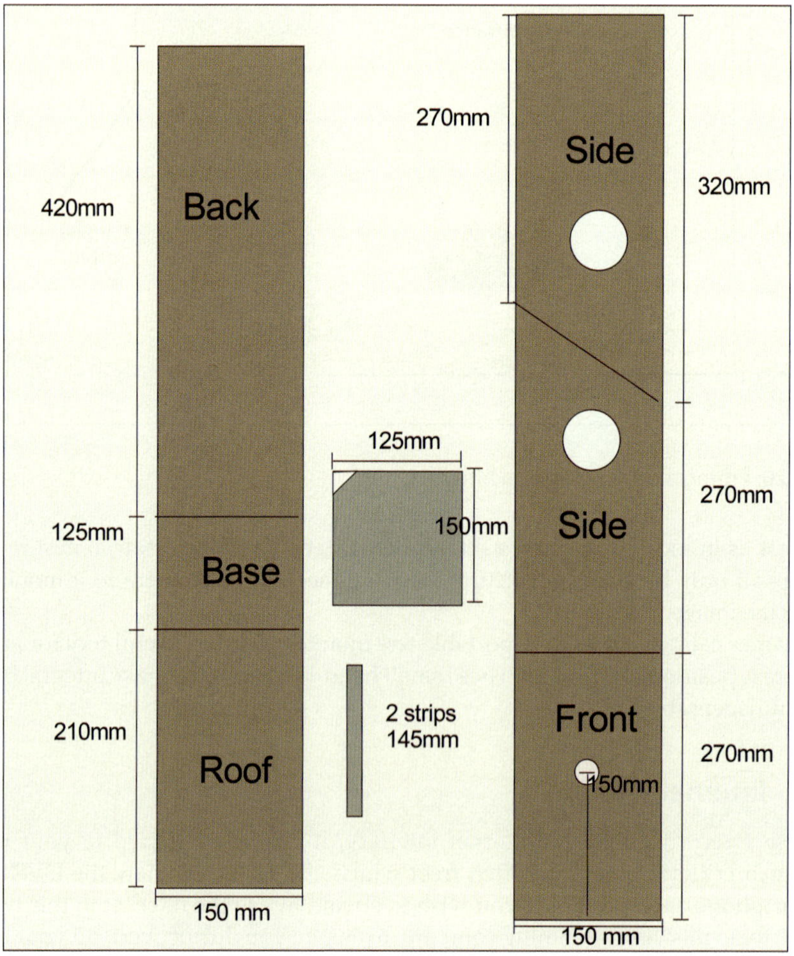

Figure 2.27 Camera nest box plan.

2.12 Bird and mammal boxes

2.12.1 Nest box plan

There are many companies that sell nest boxes complete with camera, cables, power supplies and all connectors, but it is useful to know how to build your own nest box so that you can add your own choice of camera and connectors, either from a kit or separately. That way you can adapt the box to put the camera on the side in its own box or on top looking down. The plan can be adjusted for larger birds.

This plan shown in Figure 2.27 is for a nest box made from 12.5 mm thick material, either exterior ply or wood. All materials should be untreated as treatments may be poisonous.

Step 1. Shown in Figure 2.28. Attach the strips to the sides with glue and small nails. Attach the base to the back with glue and nails or screws. Attach the sides to the base and back with glue and nails. The side holes are made so that plastic or similar can be attached to allow some light in to the box. Glue the plastic windows to the side holes with superglue.

Step 2. Cut a small groove in one side for the wires from the camera. Bolt the camera to the shelf as shown in Figure 2.26 and Figure 2.29 and place it on the strips. Leave the shelf loose so that it can easily be removed.

Step 3. Cover the roof with felt or EPDM pond liner and tack into place along the edges. Cut a piece of rubber for a hinge and tack to back and roof as shown in Figure 2.30.

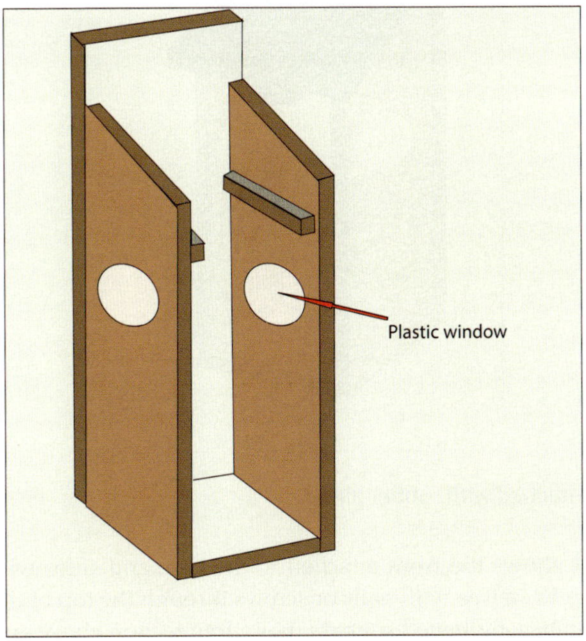

Figure 2.28 Back, sides and base fixed, with supports for shelf.

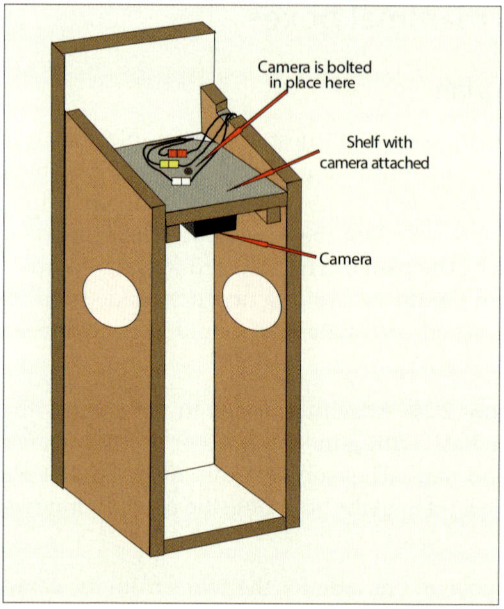

Figure 2.29 Camera fitted to shelf.

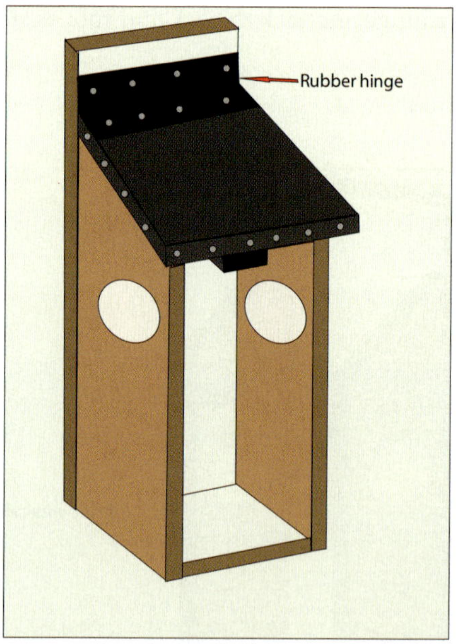

Figure 2.30 Roof attached with rubber hinge.

Step 4. Figure 2.31 shows the front attached to the base and sides with glue and nails. The box can be fixed to a tree with nails or screws through the top of the back. For larger boxes, the back can be lengthened so that it has a free section above and below the main body of the box. This allows it to be fixed top and bottom.

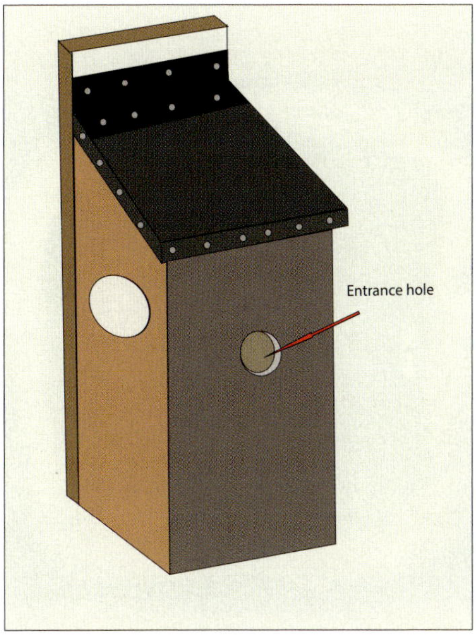

Figure 2.31 Front attached with an entrance hole.

In this box, the camera points downwards. If another hole is placed low down in the side of the box, another view can be obtained. In that case, the camera would have to be in a separate box. The box must be sited close enough to a power supply so that the resulting images can be viewed or recorded. The length of the cable from the camera will dictate the maximum distance this should be. Cables are typically 20 m long, but longer cables can be obtained.

2.12.2 Mammal boxes/feeding stations

The cameras used in bird boxes can be also be used in mammal boxes or feeding stations. The example in Figure 2.32 shows a feeding station for small mammals. A weatherproof box is all that is required, but for more attractive images, the box should be lined to give a natural appearance. Air-drying clay is good for this and can be obtained easily from the internet. I used 4 kg of reinforced terracotta-coloured clay for the box in Figure 2.32. Although this means the surfaces absorb rather than reflect light, the appearance is natural. Moss, peat, twigs and leaves can be pressed into the clay surface before it dries to enhance the effect.

The advantage of using a wooden box lined with clay, rather than clay on its own, is that it makes it easier to fix the camera to the box on the side. Also internal lighting can be added easily by attaching a small section of LED strip light to the roof.

The box is approximately 20 cm square, which gives mice and voles room to move around and still remain in focus.

Although a clay-lined wooden mammal box is attractive, it does have disadvantages. The clay becomes soft when wet and can crack when dry, and is also heavy. Details of how to make a lightweight plastic mammal box are given in section 7.4.

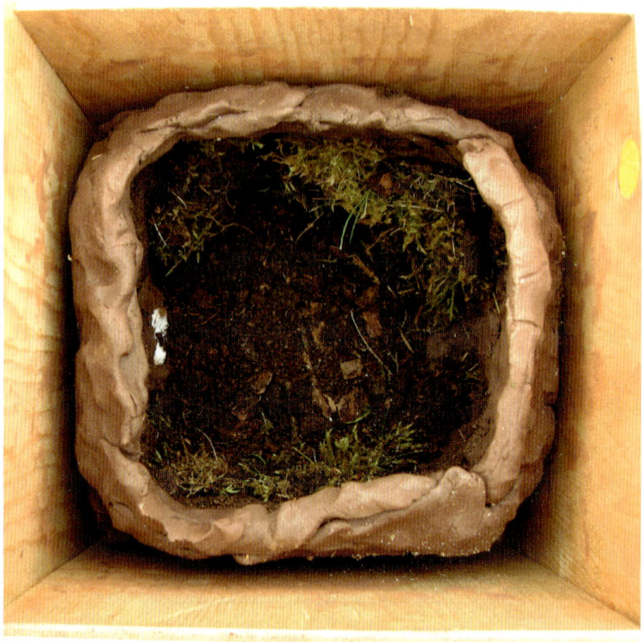

Figure 2.32 Clay-lined small mammal box.

2.13 Advanced systems

2.13.1 Wireless

The details of advanced wireless systems are not within the scope of this book, but it is important to know that such systems exist.

As described in section 2.3, the commonest frequency used by wireless CCTV systems is 2.4 GHz, with 100 m line of sight being the usual limit on distance. If used in remote areas, for example monitoring birds on sea cliffs, there is less likelihood of interference, but in other areas, interference can be a major problem. Interference can degrade the video signal and make it unusable. It can also render motion detection recording ineffective, as interference will be treated as motion by the recorder with the result that recording may be continuous.

It is possible to use different frequencies, with frequencies of 5.8 GHz or 900 MHz being used. Such systems are more expensive but have the advantage that the antennae may be able to transmit several kilometres and interference will be reduced. Care should be taken in the UK to enquire whether a license from Ofcom is required for the equipment used.

3G mobile technology can also be used by some cameras or DVRs. Success depends on there being a reliable mobile signal available where the system is to be transmitted and received. Transmission rates with 3G may be too slow for smooth live viewing except at low resolution, and the costs of transmitting large amounts of data should also be taken into account.

2.13.2 Event monitors

This book has concentrated on the motion-detection triggering aspect of recording. Some recorders and cameras also have the facility for event monitoring. This means that detectors such as PIR heat-triggered motion detectors, pressure switches and other detection devices can be used to trigger recording. The application of these techniques is more advanced, requiring some knowledge of electronics, but may be useful in complex projects.

2.14 Videos

Fox

The video in Figure 2.33 was taken with a basic analogue camera which did not have professional connectors. The video is clear but there is a colour cast. This can be removed with a video editing program.

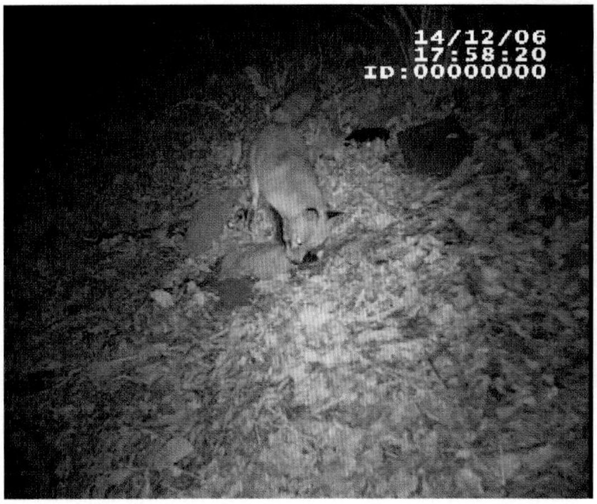

Figure 2.33 Fox taken with a basic analogue camera.
https://youtu.be/_byGfYuYNBM

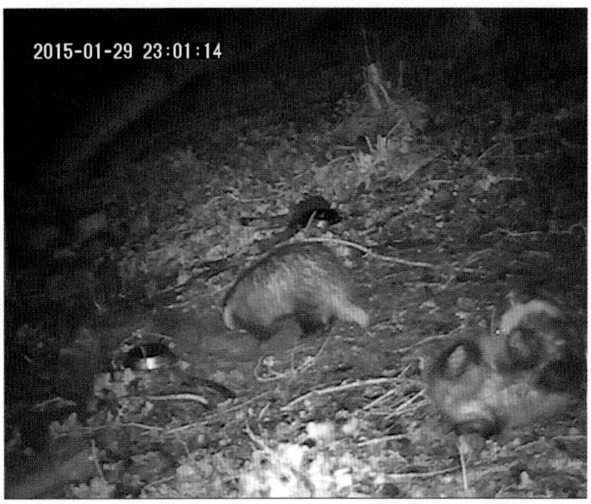

Figure 2.34 Badger taken with a basic analogue camera.
https://youtu.be/j_yiNp-28W8

Imperfect videos like this are included to show faults the novice is likely to encounter and how to improve on them.

Badger

The video in Figure 2.34 was taken with an analogue camera with professional connectors. There is no colour cast, but the video would benefit from some improvements in contrast. Video editing software is useful if image improvement is necessary.

Bat

The video in Figure 2.35 was taken with a high-resolution (1,000 TVL) analogue camera. There is sufficient detail to make it possible to identify the bat as a long-eared bat due to its hovering flying style. Its long ears are also clear. The video has good contrast and sharpness without overexposure of the tree trunk which is quite close to the camera.

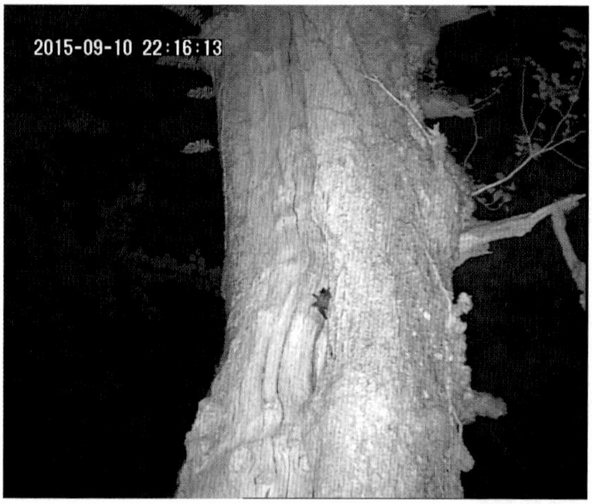

Figure 2.35 Bat taken with a high-resolution analogue camera.
https://youtu.be/U6bCPkRpG70

3. HD-TVI CCTV

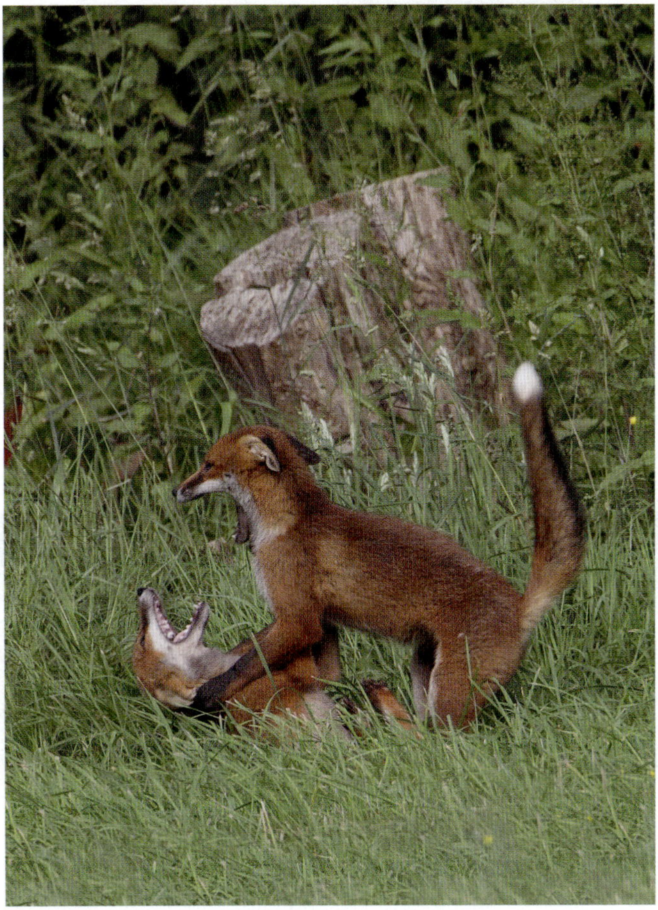

Figure 3.1 Fox (*Vulpes vulpes*) cubs make good CCTV subjects.

3.1 HD-TVI Quick Start

Dramatic improvements in quality are possible with this type of analogue HD system which uses new technology to deliver high-quality video. RG59 cables, as used for standard analogue CCTV, can be used. Apart from giving higher-quality images, HD-TVI can also be used with greater lengths of cable (up to 500 m) without signal loss.

This Quick Start assumes that a mains power supply is close at hand.

Step 1. Buy an HD-TVI camera (see Figure 3.2) such as the infrared day/night camera shown in Figure 3.2 with 1,080 pixel resolution and a lens with a variable focal length from 2.8 to 12 mm.

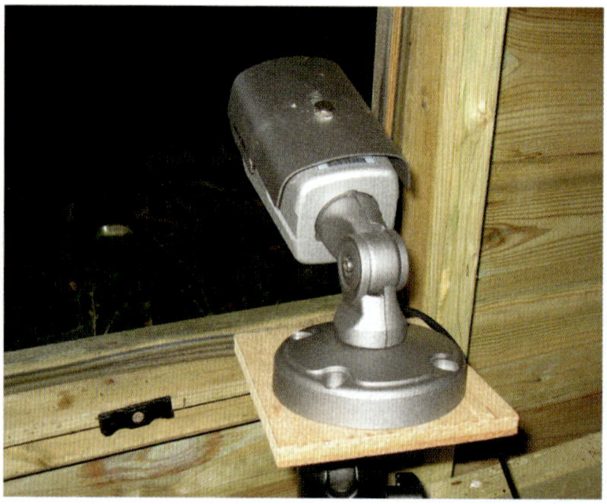

Figure 3.2 HD-TVI camera.

Step 2. Buy an HD-TVI recorder (see Figure 3.3) with a hard disc drive. HD-TVI cameras can only be used with an HD-TVI recorder, but an HD-TVI recorder can be used with standard analogue and IP cameras as well as HD-TVI.

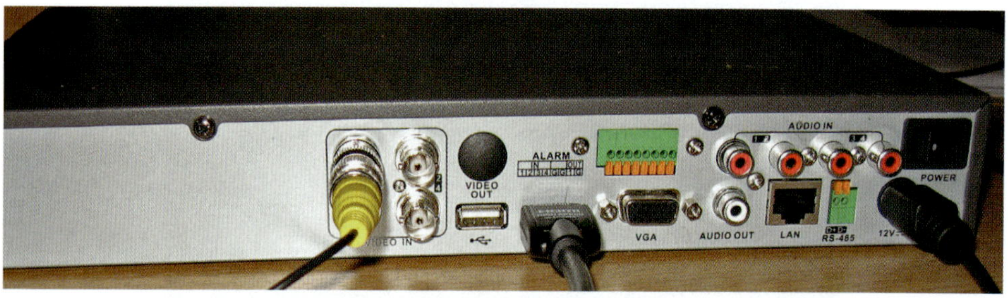

Figure 3.3 HD-TVI recorder, back view. A mouse is connected to a USB socket on the front.

Step 3. If the cable from the camera is too short to reach the recorder, use a good-quality waterproof RG59 Siamese cable (see Figure 3.4) that has wires for both video signal and power. A length of 10 m is suitable for many situations, but if the mains supply is further away, a longer cable will be required.

Step 4. Screw the camera to a wall, tree (using screws that will not damage the tree) or other suitable surface outside, where you expect wildlife such as foxes, deer or badgers to pass by. Alternatively attach to a tripod as shown in Figure 3.2.

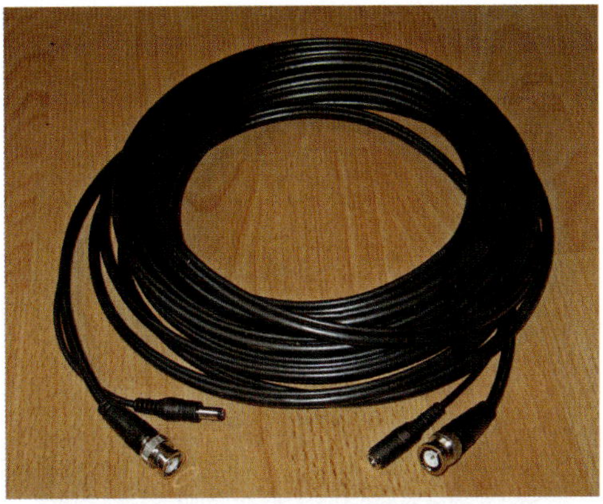

Figure 3.4 RG59 Siamese cable. Left: BNC and male power plug connect to the camera; right: BNC connects to the recorder and the female power socket to the camera power supply.

Step 5. Connect the power plug from the camera, or from the cable connected to the camera, to a 12 V DC power adapter. Plug the adapter into a mains power socket.

Step 6. Connect the recorder to a mains power socket (right-hand wire in Figure 3.3).

Step 7. Connect the (yellow) signal plug from the camera to the camera input on the DVR (see Figure 3.3). If the camera has been connected to an RG59 extension cable, the other end of the RG59 cable should be connected to the recorder instead.

Step 8. Connect a monitor to the HDMI video out connection (central wire in Figure 3.3). If the monitor does not have an HDMI input connection, a VGA cable can be used.

Step 9. Adjust the camera position so that the monitor shows the required scene.

Step 10. Use the recorder menu to select motion recording, unless you wish continuous video.

The resulting videos can either be played back on the recorder or copied to a USB drive to view on a computer. Alternatively, the action can be watched live on the monitor. An example video still is shown in Figure 3.5.

3.2 HD-TVI CCTV

3.2.1 Introduction

There is a great deal of competition worldwide in the CCTV market, and CCTV has improved dramatically as a result. New technology is continually being developed as a result of this competition and the size of the surveillance market. HD-TVI is the latest, and I believe the best, of recent advances. HD-TVI is a competitor to the quality of IP cameras while remaining analogue and straightforward to set up.

At the time of writing, HD-TVI equipment can be obtained relatively cheaply and gives very high-quality videos. The technology has not been fully adopted as yet, but as the equipment is cheaper than standard analogue equipment and gives much higher quality video, it should be the first choice for wildlife monitoring where mains power is available.

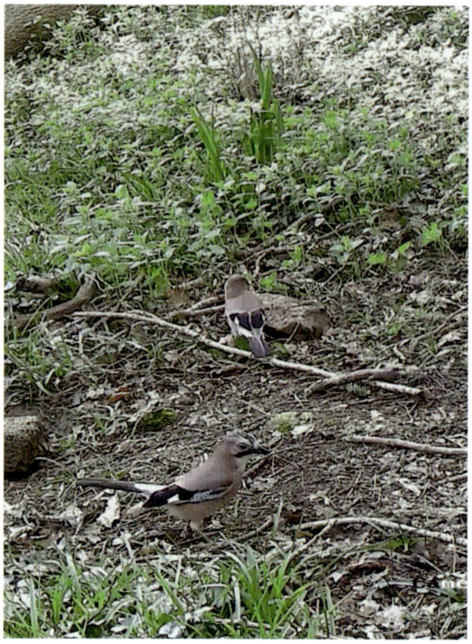

Figure 3.5 Jays (*Garrulus glandarius*) taken with an HD-TVI camera.

3.2.1.1 What is HD-TVI?

HD-TVI uses a different method to process the video signals. Although still an analogue camera, the signal quality is dramatically improved compared to standard analogue systems. Because it is an analogue and not a digital system, the recorders (which must be HD-TVI) can also process standard analogue cameras, and even IP cameras in some cases.

3.2.1.2 What does it look like?

HD-TVI systems look no different from standard analogue systems, although there is always an HDMI output from the recorder to give the HD video.

The cameras still use 12 V DC, and infrared LEDs for night viewing.

3.2.2 HD-TVI setup

Figure 3.6 shows a block diagram of a typical HD-TVI system.

3.2.2.1 Components

At the time of writing, HD-TVI has only been available for a few months, and is only now being taken up by CCTV retailers. However, the huge increase in quality with (at present) no increase in cost, plus the ability to use standard RG59 cable in long runs makes this system a winner. CCTV camera manufacturers and retailers are already cutting back on low-resolution analogue cameras because of the competition from HD analogue cameras.

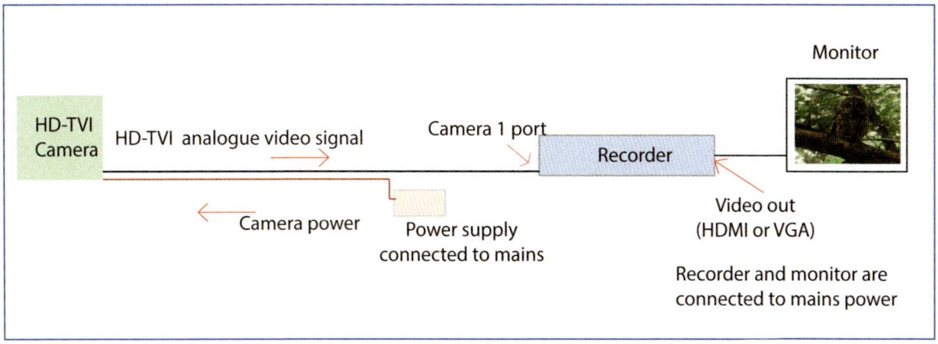

Figure 3.6 HD-TVI system.

Camera

HD-TVI cameras look very similar to standard analogue cameras and use the best features, such as 'intelligent IR', which means that objects close to the camera are not overexposed when infrared LEDs are used at night. Noise reduction and other advanced features are often included.

12 V DC is the standard power required, but the current draw tends to be larger than with standard analogue cameras. Infrared LEDs for night vision are the same, and a similar range of lenses are available. The most suitable focal lengths for wildlife monitoring are from around 2.8 to 12 mm.

Cameras with different infrared ranges are available, with 30 m being a reasonable choice, as wildlife looks rather small at distances greater than this.

Recorder

An HD-TVI recorder is required, as a standard analogue recorder will not work. However, it is possible to get HD-TVI recorders that also work with standard analogue cameras. At present the recorders all seem to have four inputs or more, and there does not as yet seem to be a small HD-TVI recorder with the input for one camera only. This means that the present systems are not yet suitable for mobile CCTV which requires a low current draw. The four input recorders draw a larger current than a single input recorder.

Because portable HD-TVI recorders are not currently available this means that HD-TVI usually requires mains power. However, some HD-TVI recorders run on 12 V DC, which means it is possible in theory to run such a recorder from a battery. The current draw for recorder and camera can be over 1 amp, with the result that batteries need frequent recharging. I have run an HD-TVI system using a lead-acid battery and a 30 W solar panel for a short period, but for longer periods a 60 W panel would be required.

Connections

The HD-TVI connections are similar to those for standard analogue cameras with a BNC from the camera and a female barrel plug for the power input. The recorders have a variety of output formats such as HDMI, VGA and RCA.

3.2.2.2 *Weatherproofing*

A range of weatherproof cameras is available, so do not buy a camera without a water-resistant rating of at least IP65. Cameras with ratings of IP66 or IP68 are fully waterproof and therefore preferable.

The recorders are not waterproof and are intended for indoor use.

3.2.2.3 *Image quality comparison*

The images below compare standard analogue with HD-TVI analogue. Analogue cameras generally have an effective resolution of about 0.5 megapixels, but HD-TVI cameras have a resolution of 2 or more megapixels.

Standard analogue CCTV images

Analogue CCTV images can be of good quality, but are small. In the infrared image in Figure 3.7, the fox on the edge of the pond is hard to see in the video still, although it is clearly seen in the video.

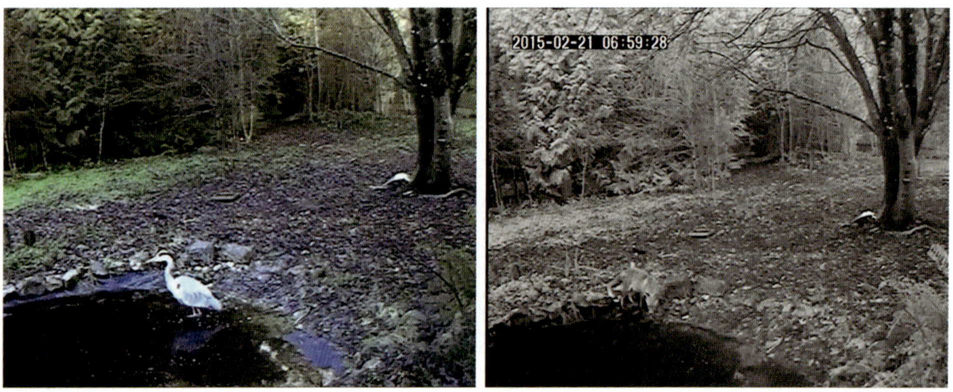

Figure 3.7 Standard analogue camera CCTV image quality.

HD-TVI image

With HD-TVI, even video stills give images of a good quality and reasonable size. There is a noticeable lack of the graininess often seen with trail camera infrared night images as shown in Figure 3.8.

3.2.2.4 *Uses of HD-TVI*

HD-TVI is the CCTV system of choice for most wildlife monitoring applications with mains power available. The range of cameras is increasing, although there are fewer very small cameras at present.

The main features of HD-TVI can be summarised as:

- Excellent-quality cheap cameras.
- Easy to set up.
- High-quality HD videos.

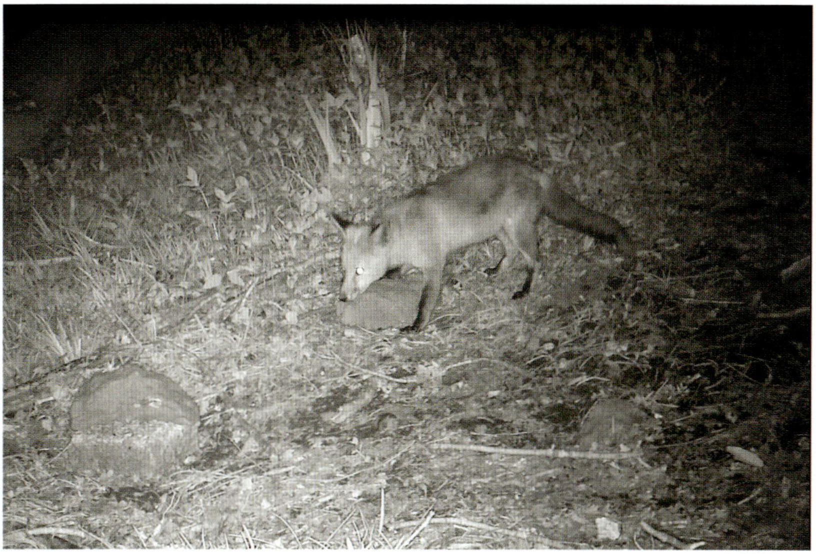

Figure 3.8 HD-TVI camera image quality.

- Long cable runs possible.
- HD-TVI recorder can also be used for standard analogue and IP cameras.

3.3 Videos

Fox cub

HD-TVI gives excellent colour saturation. The short video in Figure 3.9 shows the improvement in quality with an HD analogue camera. More HD-TVI videos can be found in Chapter 8.

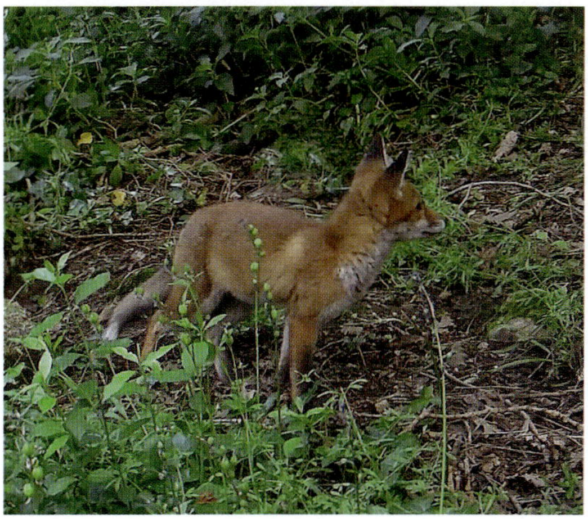

Figure 3.9 Fox cub taken with HD-TVI camera.
https://youtu.be/4AasL5VZJ00

4. IP cameras

4.1 IP camera Quick Start

IP cameras are rather more complex to set up, so this IP Quick Start is in the form of a block diagram only. Detailed step-by-step instructions for setting up an IP camera are given in section 7.3. A basic block diagram is shown in Figure 4.1.

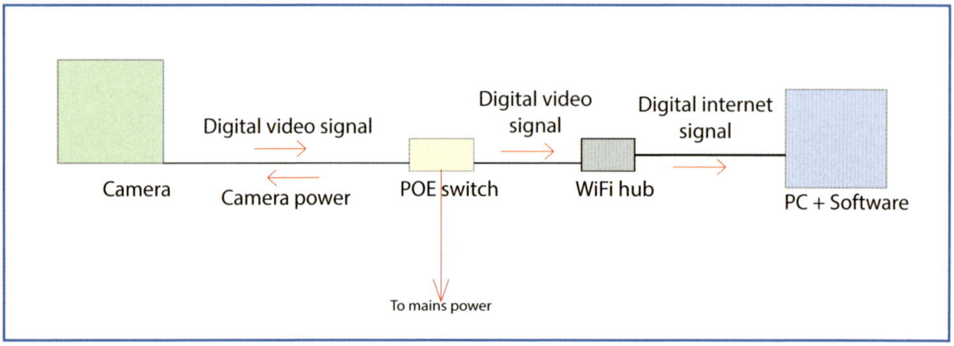

Figure 4.1 Block diagram of an IP camera system.

4.2 IP cameras (advanced CCTV)

4.2.1 Introduction

IP cameras use a technology different to that used by analogue cameras. Although the sensors may be similar to those used in analogue cameras, the electronic circuits in the camera convert the analogue signal to a digital signal and make use of network technology to carry the video data. This can produce a high-quality video if there is a fast internet service.

The cameras are more expensive than analogue cameras but until the advent of HD analogue systems were the best method to obtain high-quality videos, with the advantage that they can be accessed remotely. Wildlife videos, such as the Norwich Cathedral Peregrine Live Web Cam, are often streamed to the internet and are a popular way to ensure a wide viewing audience.

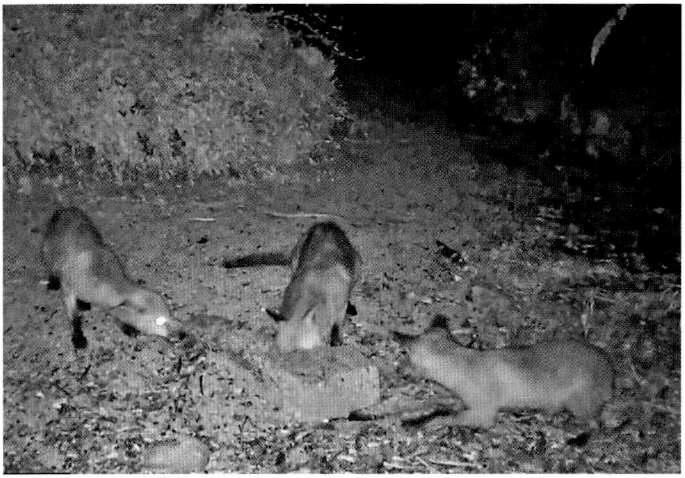

Figure 4.2 Still from an IP camera video.

As there are many similarities with analogue cameras, I have concentrated on those areas that are unique to IP cameras. An example still is shown in Figure 4.2.

4.2.1.1 *What are IP cameras?*

IP stands for Internet Protocol. IP cameras receive an analogue signal from the sensor and transmit this as a digital signal along an ethernet (Cat5) cable to a router using computer network technology. The camera data can be viewed on a computer connected to the router either physically or by WiFi. IP cameras are usually able to get their power through the wires in the cable with the use of an additional Power Over the Ethernet (POE) switch.

4.2.1.2 *What do they look like?*

IP cameras are similar in appearance to other CCTV cameras, and can supplied with or without weatherproof housings. Most have infrared LEDs for night vision. IP cameras are generally regarded as technically superior CCTV cameras and have many advanced features.

For variable-focus cameras, zoom and focus can be adjusted manually using screws on the camera body. The more expensive models offer digital zoom, which is a very useful feature as it can be controlled remotely.

4.2.1.3 *Uses of IP cameras*

The uses of IP cameras are similar to analogue cameras, except for the fact that both mains power and an internet router are necessary. IP cameras are available in small sizes, but to take advantage of the higher image quality required for wildlife, larger sensors and thus slightly larger cameras are more suitable. Small IP cameras can be used in bird boxes and mammal boxes, and the larger cameras can be used for external monitoring of larger mammals and birds.

4.2.1.4 *Weatherproofing*

IP cameras can be obtained in 'indoor' quality which means they must be protected from rain or high-humidity environments. However IP cameras can also be obtained in weather-sealed housings, although at greater expense. As with analogue cameras, a water-resistant rating of IP65 to IP68 is advised. Do not confuse the weatherproofing IP rating with the acronym IP for Internet Protocol cameras.

4.2.1.5 *Power*

Many IP cameras are POE compliant. This means that they do not require a separate power supply, but rather get their power through the ethernet cable carrying the camera signal. The other end of the cable is connected through a POE switch to a router. The power for the camera is supplied from the POE switch (which is plugged into the mains) and the video signal from the camera is directed through the POE switch to the router.

4.2.1.6 *Cost*

Cost varies according to resolution, whether the camera is waterproof, whether it has night vision (infrared LEDs), the range of the LEDs, and whether the lens has fixed or variable focus. It is possible to obtain cheap unbranded cameras, but I would suggest it wiser to buy cameras from companies with a proven track record of reliability. This is true for most CCTV equipment but is particularly so for IP cameras which are complex to setup, and often require the advice of a friendly supplier. The cost of a good IP camera for wildlife use is around £200–400.

4.2.1.7 *Examples of use*

Colour videos by day

IP cameras generally give high-quality images in daylight provided the internet service is fast enough.

Infrared videos by night

Most outdoor IP cameras have built-in infrared illumination, but others may need a separate light source. The image quality degrades in infrared light, but this does depend on the camera make, and the power of the illuminators that are used.

The main features of IP cameras can be summarised as:

- Image quality good to very good.
- Can be viewed live via the internet.
- Can be complex to set up requiring some basic knowledge of internet technology.
- Most expensive form of CCTV.
- Requires a reliable fast broadband connection.

4.3 IP cameras in depth

4.3.1 Technical details

The basic structure of an IP camera is shown in Figure 4.3.

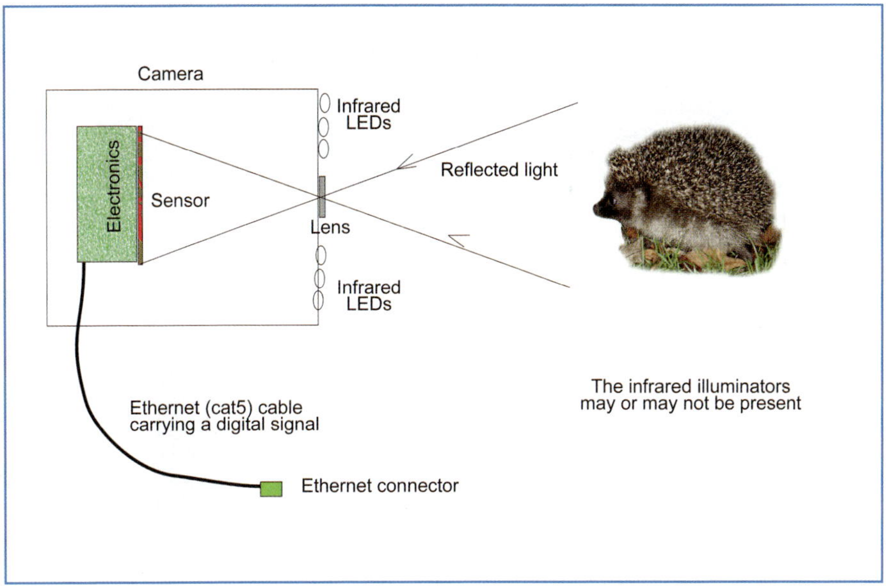

Figure 4.3 IP camera detail.

The length of an ethernet cable for use with IP cameras is limited to 100 m without boosters. The same challenge of dealing with cables applies to IP CCTV cameras as with other cameras. An additional challenge is that the cable must connect to an internet router/hub, which may not be conveniently located nearby.

4.3.1.1 POE switches

The ethernet cable from the camera plugs into a POE switch (see Figure 4.4 and Figure 4.5) which supplies power *to* the camera and transfers data *from* the camera. The POE switches come in various sizes, typically taking 1–4 cameras, but larger, and correspondingly more expensive, switches can take many more cameras. The switches have to be plugged into the mains to supply the necessary power to the cameras.

4.3.1.2 Home plugs

If you have WiFi but your WiFi signal does not reach to where the ethernet cable comes into the house from the external camera, you can use home plugs as shown in Figure 4.6. These plugs pass the WiFi signal along the mains electricity wires. This means that long Cat5 cables are not required inside the house.

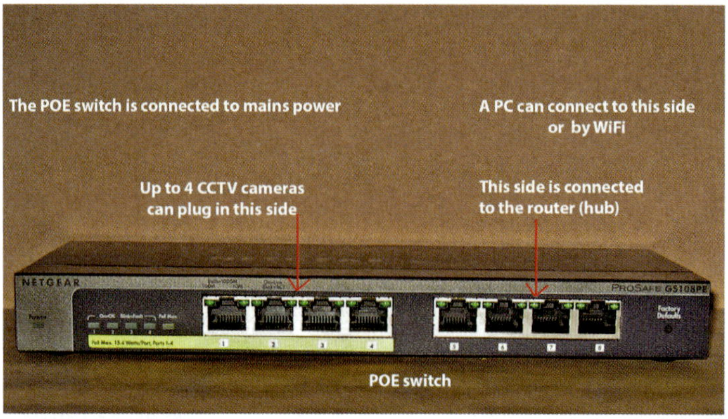

Figure 4.4 POE switch.

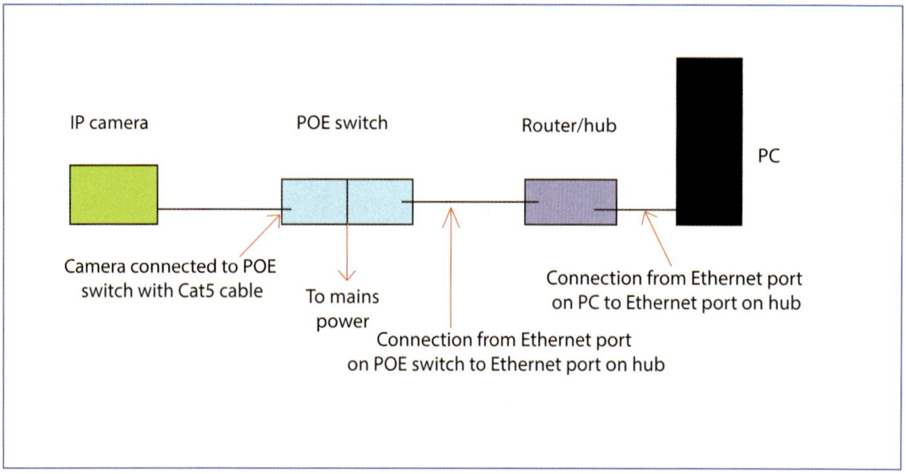

Figure 4.5 Connecting an IP camera via a POE switch.

The home plug connections are shown in Figure 4.7. The first home plug is plugged into a mains socket close to where the Cat5 cable comes from the camera. The Cat5 cable is plugged into the home plug, thus connecting the camera to the home plug. This first home plug must have an integrated POE switch.

The second home plug is plugged into a mains socket close to the broadband router/hub. Provided both home plugs are on the same mains circuit, the signal from the camera will pass through home plug 1, through the mains wires , through home plug 2, into the router and to the PC as an internet signal. The signal may be degraded to some extent, but the convenience of home plugs may outway this disadvantage.

4.3.1.3 Battery life

Generally speaking, batteries are not used with IP cameras, as power is supplied via the ethernet cable. This is an advantage if both mains power and the internet are available, but it means that IP cameras cannot be used in portable systems.

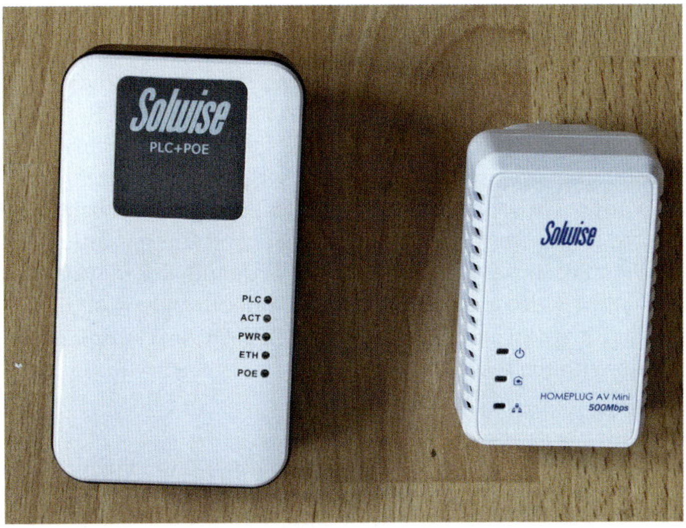

Figure 4.6 Home plugs. The left-hand plug has an integrated POE switch.

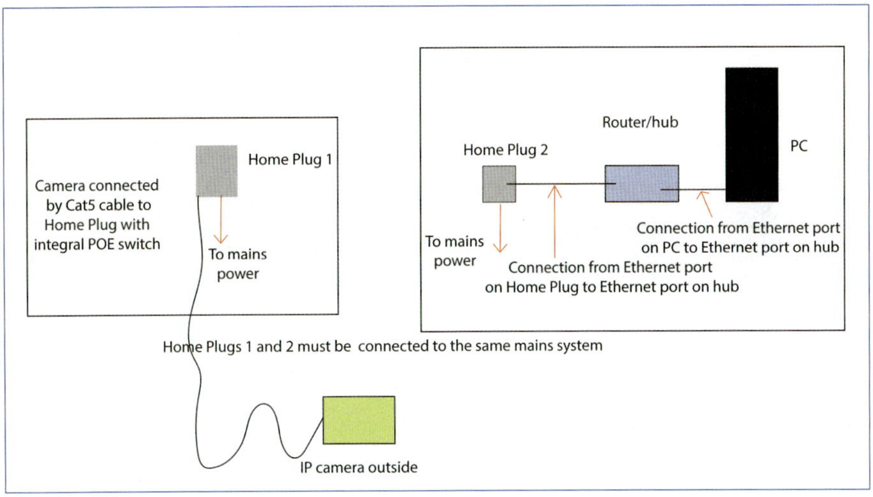

Figure 4.7 Using home plugs to extend internet reach.

4.3.1.4 Image quality

IP cameras transfer digital signals directly from the camera to the PC (or strictly speaking to software on the PC) via the internet. The images are of much higher quality than the equivalent CCTV analogue cameras.

Sensor resolution

The resolution of the sensor varies according to the camera model, but is commonly 1.3 megapixels. More expensive models have a resolution of 2 megapixels of more.

Video size

IP cameras can be used without a digital video recorder and are often used for live streaming to the internet. The software that comes supplied with the camera, or separate software such as iCatcher, allows the recording of videos on a PC.

Because IP cameras produce high-quality HD images, a large amount of data has to be carried by the internet system. This is often described by saying that the camera takes up a large bandwidth. Those who have a slow to average broadband speed (less than 10 Mbps) will have to reduce the video screen size, and the resolution of the image, to prevent jerky videos. For those with high-speed broadband, this may be less of a problem.

The adjustment of video parameters to give smooth videos of good quality is not straightforward and if you wish to buy an IP camera, chose a supplier who will be able and willing to help with the setup.

Light sources

IP cameras usually have built-in LED lights for night time or for viewing in low light levels, but the cheaper models may need a separate infrared light source. The disadvantage of a separate light source is that power must be supplied, which usually means an extra cable is required with the ethernet cable. If you make up your own cables, unused wires from the Cat5 cable can be used to carry power to the light source.

When cameras are used in confined spaces such as mammal boxes, self-adhesive LED light strips are a very useful form of illumination (see Figure 4.8). Short strips can be cut off and attached to the roof of the mammal box. LED strips require a power source, typically 12V DC, but as their current drain is low, batteries can be used as a convenient alternative to a separate power cable.

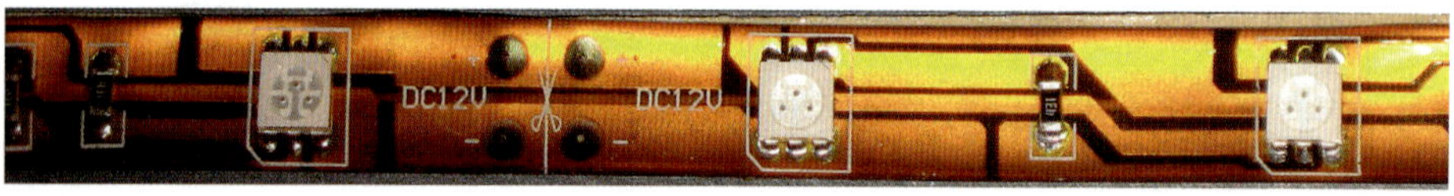

Figure 4.8 LED light strip.

It is possible to purchase short LED light strips with wires for connecting to a power source already incorporated, but it is more likely that wires will have to be soldered to the contacts. As the most useful LED strips have a thick waterproof coating, this must first be removed. Instructions for this are usually given by suppliers.

It is also possible to obtain infrared LED strips for night filming in a mammal box, but these are more difficult to find in the UK. This is one area where eBay may be useful, and it is possible to find reliable foreign suppliers using eBay.

4.3.1.5 Motion detection

One advantage of IP CCTV cameras is that the software that brings the camera image to the screen can often be set up to record by motion detection only. This form of motion detection is similar to that used by recorders, i.e. the actual motion of groups of pixels will trigger recording.

4.3.1.6 Range and bandwidth

The range of an IP CCTV camera will depend on the camera model and in particular the power of its integrated infrared LEDs as with an analogue camera. As bandwidth is an issue with IP cameras, a camera that gives a clear image of subjects 30 m away is not necessarily an advantage, and reducing the resolution of the image for smooth videos will be less obvious if the camera covers a smaller area, as there will be less detail to transmit. For this reason, cameras with a variable-focus lens are useful, as this makes it easier to zoom in on a subject and cut out unwanted backgrounds that would otherwise increase bandwidth.

4.3.1.7 Field of view

As with standard CCTV cameras, IP cameras vary a great deal in field of view, depending on the lens used. A small focal length of 2.8 mm will give a wide field of view and a greater depth of field, meaning that more of the image will be in focus.

A wide field of view can be a big disadvantage if filming using a mammal box. In Figure 4.9 the wide field of view shows the LED strip lights in the roof of the box. Adding a moss covered lid and pushing the LED strips further back would solve the problem.

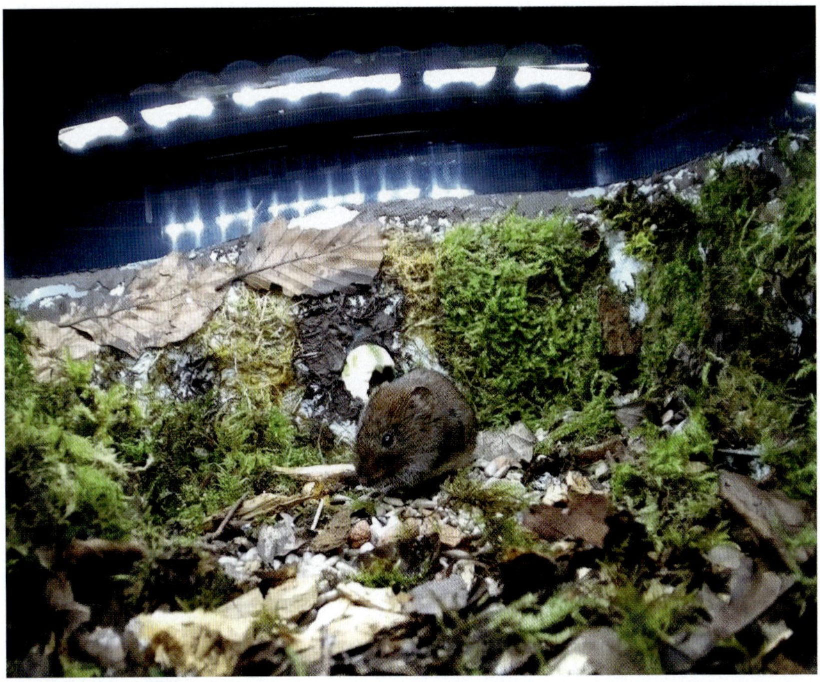

Figure 4.9 The effect of a wide field of view in a mammal box.

4.3.1.8 Camera mode

Video

IP cameras are used to produce videos, but because the images are of such high quality, it is feasible to take stills from the videos produced, and process them using propriety software such as Photoshop.

Day/night

As with standard CCTV cameras, IP cameras often have integral infrared LEDs, usually positioned round the lens. The same problems of false colour can be found as a result of the sensitivity of the camera sensor to both visible and infrared light. The better-quality cameras have an infrared filter than is positioned in front of the sensor during the day, and which is automatically removed when light levels fall and the inbuilt infrared LEDs supply illumination. This is often called a 'cut' filter, and such cameras are called 'true day/night' cameras.

4.3.2 Setting up

4.3.2.1 Outdoor

An example of an IP camera setup is shown in Figure 4.10. The most convenient method for fixing IP cameras is to screw the camera base to a piece of wood and attach the wood to a post or tripod. The additional electronics in a high-resolution IP camera can make it heavy, and this method makes it much easier to remove the camera and reposition it if required.

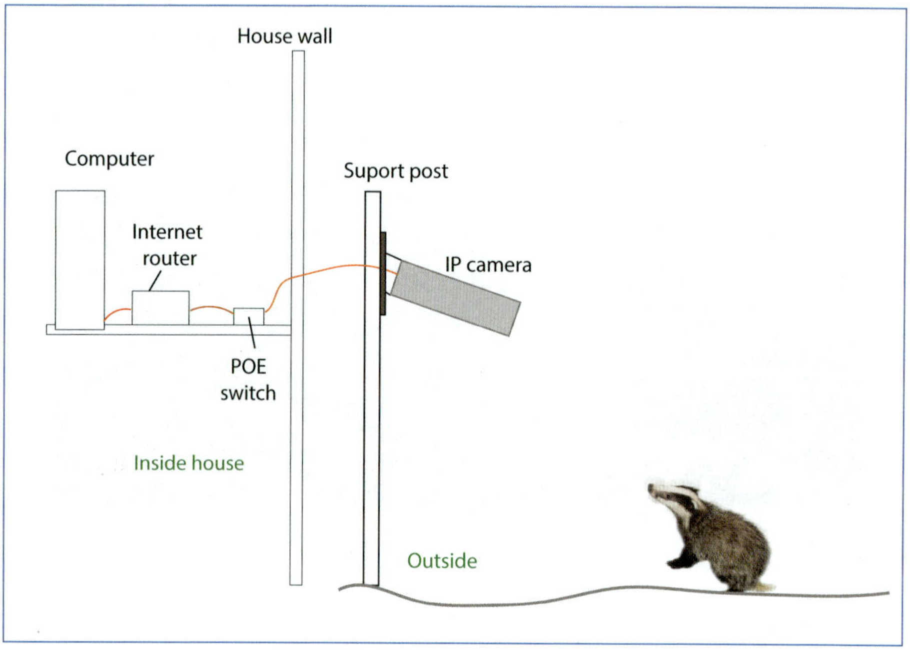

Figure 4.10 IP camera setup.

4.3.2.2 Indoor

A mammal box is one example of an 'indoor' application. The boxes can be positioned outside, but the camera must be positioned inside a container and thus protected from wet.

The details of how to make a mammal box can be found in section 7.4. If the inner and outer plastic box method is used, a wooden stand may be required to give a suitable support for the camera and to allow accurate positioning within the box.

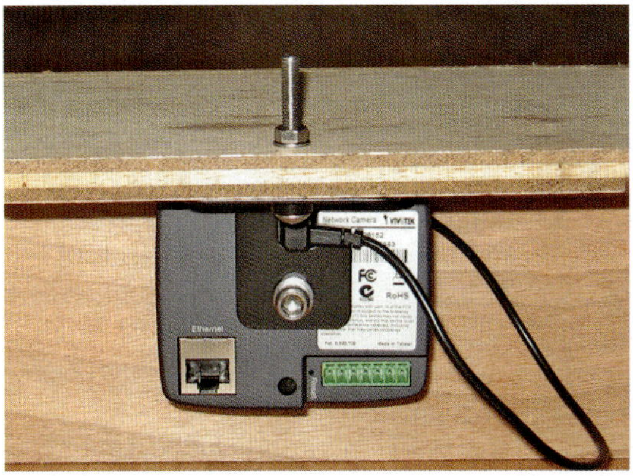

Figure 4.11 Stand for a small IP camera.

Figure 4.11 shows the back view of a Vivotek 8152 camera bolted to a wooden stand which has three sides (for stability) and a top (for bolting the camera to). A stand such as this separates the camera from the mammal box visitors, which is a necessary precaution as small mammals invariably jump out of the inner box if it does not have a lid and are liable to chew the cables.

4.3.2.3 Security

As the feed from IP cameras is relayed live to a PC (or other device), it is easier to protect the camera from theft, as there will be a permanent record of any intruders. If the camera is switched off there will be no live feed, but an intruder would not be able to tell whether the camera was on or not, and would hopefully be deterred.

4.3.3 Working with images

4.3.3.1 Recording images

A big advantage of IP cameras is the availability of computer software for recording. Software such as iCatcher has a wide range of features to make recording very flexible and superior to that with a DVR. The software also allows sophisticated masking of moving objects such as foliage, which means that its movement will not trigger motion detection recording.

Recordings can be selected by time and exported to a folder on the PC.

4.3.3.2 Viewing images

Viewing on a PC

If software such as iCatcher is used, it is easy to view the live feed, view a recording or view both at the same time on a PC. The software must first be installed on the PC receiving the video signal from the camera.

Remote viewing

As stated above, a big advantage of IP cameras is that remote viewing is possible. This is because each IP camera has an internet address which can be accessed directly. The feeds can be viewed on most smartphones and tablets, either directly using the browser or by using a smartphone app which makes viewing simple.

Although this is a great advantage, its execution is not easy. Setting up the network parameters is a specialised task, especially as efficient use of bandwidth is paramount. It is advisable to use a streaming service to avoid setup difficulties and to ensure that bandwidth is protected. Streaming services mainly deal with IP cameras so are likely to be compatible with popular IP brands. If you plan to stream videos via the internet for remote viewing, make sure that the camera you choose will be supported.

Analogue camera videos can also be streamed to the internet, but only if the recorder they use has an embedded web server.

The cost of streaming services varies according to the desired quality, frame rate and pixel dimensions of the video to be streamed as well as the likely number of viewers at any time.

4.4 Videos

Roe buck

Th IP video of a roe buck shown in Figure 4.12 is a little jerky. This is because my broadband was not fast enough to transmit the data at the frame rate I wanted. The usual frame rate is 25 frames per second (fps). This video transmitted at 3.5 fps. Reducing the resolution of the video by reducing its pixel dimensions would solve such a problem, but means that much of the resolution of which the camera is capable is not used.

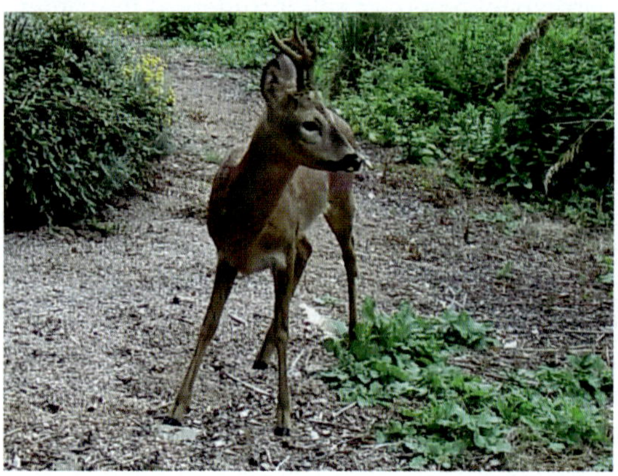

Figure 4.12 Roe buck taken with an IP camera.
https://youtu.be/4uNsy04XtZ8

Shrew

The IP video in Figure 4.13 gives an indication of the quality of video that IP cameras are capable of when the broadband speed is in sync with the video data transfer. The mammal house shows a very energetic shrew *(Sorex araneus)* rushing around collecting food. Shrews have a high heart rate and seem to live at a rate twice as fast as anything else. This IP video is smooth because the field of view was small, being in a confined space; as a result there was less information to be processed and my broadband could transmit the video at 25 fps.

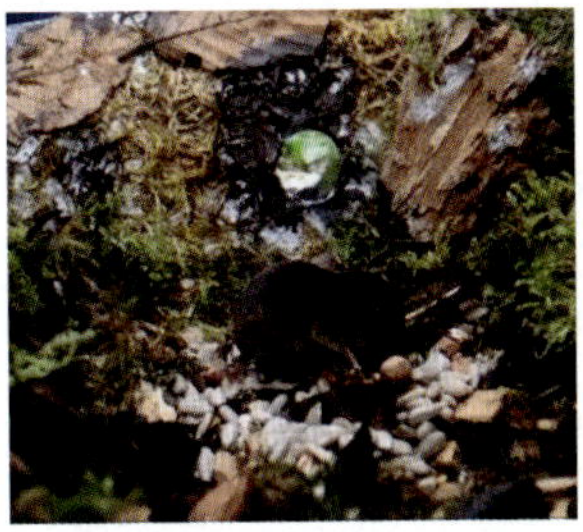

Figure 4.13 An energetic shrew.
https://youtu.be/rlpcfcHmB4Y

5. Portable CCTV

5.1 Portable CCTV Quick Start

There is no Quick Start for portable CCTV due to its specialised nature. Section 7.2 has detailed step-by-step instructions for setting up a portable CCTV system.

5.2 Portable CCTV

Portable CCTV can cover similar territory to trail cameras, as described in the CCTV overview at the start of the book. This chapter is for those who think portable CCTV would be more suitable for their application.

5.2.1 Introduction

CCTV is a well-known technology, although rather mysterious to those new to it. Portable CCTV is no more complex and is often the only practical way for using CCTV in the wild as many wildlife areas have no power, no phone or internet signal and may be several miles from the nearest habitation.

Although the components are similar to standard CCTV: camera, power source, recorder and monitor for setting up, adding the mobility aspect is far from straightforward to those starting from scratch, especially as its secrets are hard to unravel. The problem lies not in the system, but in obtaining suitable components. However, portable CCTV is straightforward to implement once these components have been identified.

Standard analogue CCTV cameras are more suitable than other camera types, and are available from many reliable manufacturers. Portable recorders are more difficult to find, but there is a range of portable recorders, priced both for the enthusiast and the professional wildlife surveyor.

Many portable recorders use SD cards, storing up to 32 gigabytes (GB), which is usually sufficient for several days of recording. Those that can record for longer have removable hard drives, but these draw a larger current and more powerful batteries must be used.

5.2.2 Technical details

The components for portable CCTV are as follows:

- Camera.

- Recorder.
- Batteries.
- Image transmission components.
- Waterproofing.
  - Box.
  - Power connections.
  - Video data connections.
- Security features.
- Portable test monitor.

A possible portable CCTV system is shown in Figure 5.1.

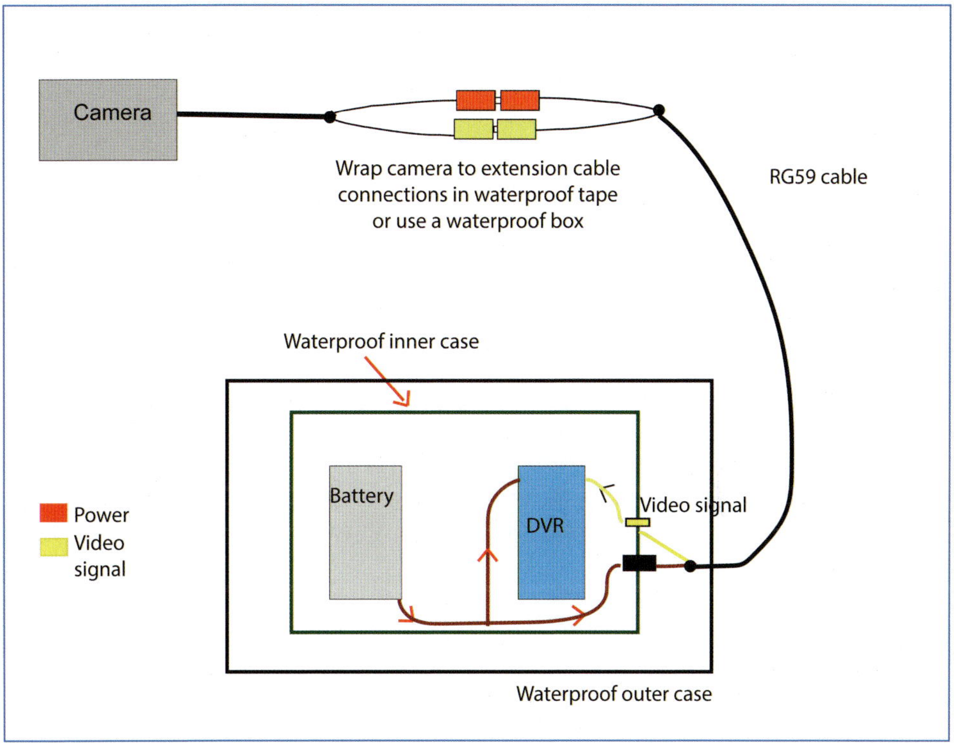

Figure 5.1 Portable CCTV setup.

5.2.2.1 Cameras

There is a wide range of cameras available for CCTV, although a more careful choice is required for portable CCTV. One must consider various points:

Is the target species nocturnal or diurnal?

For nocturnal animals, cameras with high infrared sensitivity are necessary. As white light powerful enough to illuminate the scene would require a large proportion of the battery power, infrared LED lighting should be used, especially as white light may disturb the target species. The most suitable CCTV camera would be one with integral infrared LEDs, giving a black and white image. Some cameras can prevent overexposure from the reflected infrared (smart IR), and may also have noise reduction, which makes for a cleaner image.

For diurnal wildlife, natural visible light is best (sunlight) and the image formed will be in colour. For wildlife out in the open, there is usually sufficient light to give a clear colour image, whereas in woodland, especially in summer when the leaves are out, light levels are low. Despite this, many analogue cameras are sensitive to very low light levels and give bright colour images even under low-light conditions.

Will integrated or separate illumination be required?

The integrated LEDs that come with many CCTV cameras make illuminating the night scene straightforward, especially as cameras with different LED power can be obtained, giving ranges from a few metres to 50 metres or more, depending on the camera.

In some cases, integrated illumination round the camera lens is not a good idea. Objects close to the camera may reflect so much infrared that the image is overexposed. The better cameras correct for this to some extent using 'smart IR', and careful positioning can often avoid overexposure.

However, when integral LEDs are a problem, a camera without LEDs should be used with a separate illuminator positioned so that overexposure is not an issue. This illuminator may be an extra drain on the battery.

Is the camera intended for amateur or professional recording?

Many CCTV cameras are suitable for both amateur wildlife recording and professional surveillance recording, but there are differences. As described previously, cable and connector quality can have a major effect on image quality. If you are considering portable CCTV, high-quality cables and connectors are essential.

Professional cameras may also have many features that are designed to help improve image quality, but which also make the camera more complex and hence more expensive. Make sure you are not paying for features you will not need. The more advanced cameras may also draw a larger current.

Suitable cameras I have used are:

- Twilight 2.8–11 mm variable focus, 600 TVL and 1,000 TVL, 30 m range, IR analogue camera.
- Genie 2.8–11 mm variable focus, 600 TVL and 1,000 TVL, 30 m range, IR analogue camera.

When I started this book, cameras with resolutions of 450–700 TVL were the norm, but several months later, lower-resolution cameras were scarce, and 1,000 TVL was the norm. CCTV technology is advancing rapidly.

What connectors does the camera have?

Generally speaking, 'wildlife' cameras have RCA connectors and 'professional surveillance' cameras have BNC connections for the video signal, but there are many other types of connectors that can be used. In addition, the connectors carrying power to the camera can vary a great deal. Portable CCTV requires a good knowledge of the different types of connectors.

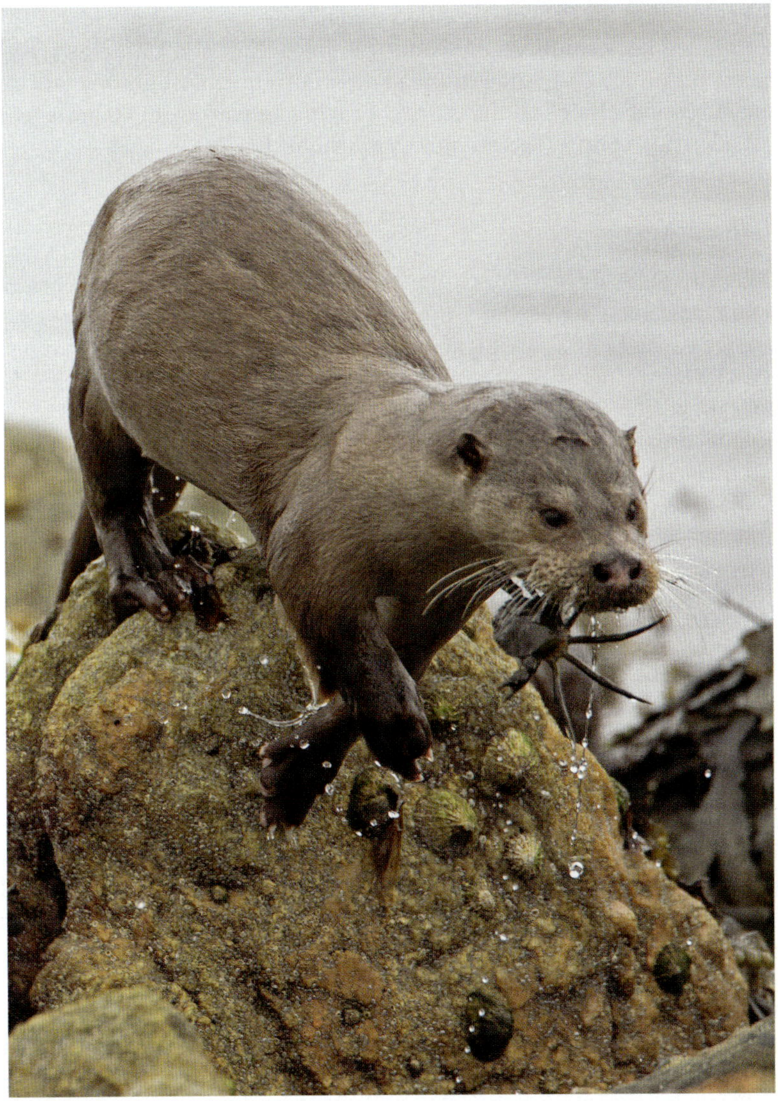

Figure 5.2 The Eurasian otter (*Lutra lutra*) is diurnal and would make a good portable CCTV subject.

Will the camera work with a 12 V DC battery?

Many of the components used in CCTV use 12 V DC. If mains power is used, the connection usually consists of a mains to 12 V DC adapter. However, not all cameras run on 12 V DC and not all run with a battery, so it is important to establish this before buying a camera.

What current will the camera draw?

This is a vital piece of information. Successful portable CCTV depends on as low a current drawn as possible to reduce battery cost and the time spent switching batteries. A total current below 700 mA (including integrated LEDs) is advisable.

What is the resolution (in TVL) of the video signal?
The higher the signal resolution, the better the image quality. Cameras with a higher resolution, such as 1,000 TVL, may draw a larger current, although this will depend on the manufacturer and model, as the current drawn can vary from 400 mA (suitable for portable CCTV) to 1,000 mA (less suitable for portable CCTV).

5.2.2.2 *Recorders*

The vast majority of DVRs used for CCTV are large, bulky, draw a large current and are unsuitable for portable CCTV. Many portable 'spy' DVRs tend to be very small with a limited storage capacity of a few gigabytes and a battery life between charges of a few hours only. However, there are several suitable low-cost DVRs for portable CCTV, such as the Genie SD-DVR.

Portable recorders are not always intended for damp or dusty environments, but I have found that a double box system seems to offer sufficient protection.

Portable DVRs produced for the army and the police are 'ruggedised' to make them less vulnerable to knocks and are suitable for harsh environments. They are also much more expensive, and possibly more suitable for the professional.

When considering DVRs suitable for portable CCTV, one must consider the following:

Will the recorder work with 12 V DC?
Most CCTV cameras require 12 V DC and as the camera and the recorder in a portable system use the same battery, recorders must have the same voltage requirements as the camera.

What maximum current does it draw (with one camera attached)?
This is vital information. Ideally, DVRs should draw less than 500 mA so that the battery power required is as low as possible. The SD-DVR draws a current of 150 mA. However, if a DVR with a hard disk is used for greater storage, the current drawn is likely to exceed this figure and sufficient battery power must be made available.

Does it have a range of recording qualities?
Most portable recorders allow a range of resolution and frame speed settings. At low resolution the SD card will last longer before filling up, but reducing video resolution may have little effect on the current drawn from the battery, and will degrade video quality. It is vital to test any equipment beforehand so that it is clear how long an SD card will last and the number of ampere hours used by the system.

Does it have motion detection?
Many wildlife applications depend on motion detection, so this is a vital feature for a portable DVR. However, one must remember that it is true motion rather than a change in heat that is detected. Masking, if available, may be required to prevent areas with moving foliage or water from giving false triggers.

Does it have pre-record facility?
This is a very useful feature for fast-moving wildlife, and is a great advantage of CCTV. As the camera is running continually (with recording being triggered by motion) it is possible for the DVR to store a few seconds of image in memory before moving on to the

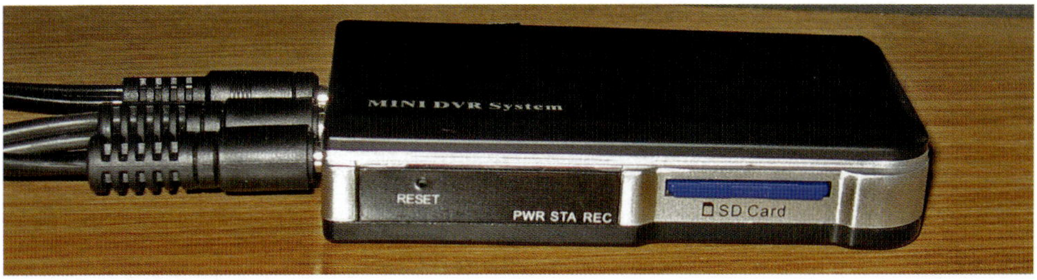

Figure 5.3 The Genie portable DVR.

next group of images. This means that if an event is triggered, the 5–10 seconds before the event will also be available and the DVR will not miss any motion. This avoids the problems of slow detection and triggering found with trail cameras. An example of a suitable portable DVR is shown in Figure 5.3.

Is the record time after triggering variable?

In some situations, for example monitoring fish, it is helpful to have a recording last for a short time, otherwise there will be several seconds (or more) of emptiness once the fish has swum past. If a DVR has the ability to adjust the recording time after being triggered, this can remove a great deal of time-consuming 'dead space'.

Is the motion-detect sensitivity variable?

This is an important feature and can help alleviate the problem of false triggers. By making the motion detection less sensitive, one is in effect allowing only changes in large blocks of pixels to trigger recording. It is important to experiment with this feature and match the sensitivity to the size of the target species. For example, deer would require low sensitivity, whereas mice would require high sensitivity.

What connectors does it allow?

CCTV has a wide range of connectors: BNC male, BNC female, RCA male, RCA female and others. Matching camera output and monitor input to the DVR connections usually requires adapters, especially from BNC to RCA, but these are readily available. The best-quality images will be achieved with the minimum number of adapters and the shortest cable lengths.

5.2.2.3 Batteries

Power is the main challenge for portable CCTV and you should consider carefully how long you wish the system to record without intervention. The shorter this time is, the easier it will be to design a portable CCTV system. Recording for periods greater than a week may require recorders with detachable hard drives. Such recorders draw a larger current and so require higher-capacity batteries.

Daily visits

For daily checks of recordings, a battery with a capacity of between 8 and 12 Ah is suitable, providing your system does not draw a large current. Two batteries will be required so that the depleted battery can be replaced and recharged. However, if recording only is required for part of a 24 hour period, a single battery could be recharged during the rest of that period.

The cheapest option is a sealed lead-acid battery. Batteries with a low capacity rating, are not too heavy and are often used with trail cameras. The higher the rating required, the heavier the battery. For example, a 45 Ah sealed lead-acid battery weighs around 15 kg and a 125 Ah battery around 30 kg.

The most convenient battery type is lithium. These are readily available, but one should be wary of unbranded batteries, as cheap lithium batteries can catch fire when charging. Named brands only should be used. One of the best (I believe) is Tracer; their batteries are light, easy and safe to handle, and pleasant to use. For low ampere hour ratings, Tracer batteries cost less than £100, but for higher ratings, such as 80 Ah, the cost can be well over £1,000. The weight of an 80 Ah lithium battery is around 10 kg.

Longer-term recording

If you wish to record for days at a time, sealed lead-acid batteries with a large capacity will be the cheaper option by far, but they are very heavy and will require a trolley or similar to transport them where there is no vehicular access. In addition, lead-acid battery voltage tends to fall as the battery discharges, so that usable capacity is only 60% of total.

Thus a nominal 60 Ah battery will only supply a usable 36 Ah. It is possible to obtain lithium batteries of a similar capacity but they cost five times as much. However, the discharge voltage is almost flat, which means that an 8 Ah battery will supply a full 8 Ah.

I have used two battery types for monitoring. A 22 Ah lithium polymer battery which lasted 2–3 days and an 80 Ah Lithium battery which lasted 5–9 days, depending on the camera used. A higher capacity battery means fewer visits between charging.

Power, current and voltage

For those whose school science is rusty, the battery that you use supplies the voltage (I have assumed 12 V DC). The appliances attached (camera and recorder) will draw a current which depends on their internal electronic makeup. For a fixed voltage, this cannot be changed. Batteries are rated in ampere hours (Ah). Thus an 8 Ah battery will supply 1 amp for 8 hours (or 8 amps for 1 hour, 2 amps for 4 hours, etc.). This is why it is very important to be aware of the current drawn by your camera and recorder. Some just quote power, in watts (W). In this case, current in amps is power in watts divided by the supplied voltage, so a 12 V DC camera rated at 6 W will draw 0.5 A or 500 mA.

Solar panels

One solution to battery depletion is to connect a small solar panel. Battery, solar panel and load (camera and recorder) are best connected using a charge controller. This prevents overcharging of the battery and gives an indication of the battery voltage remaining. If there is sufficient light, the solar panel will trickle-charge the battery, ensuring that it stays charged for longer. Panels can be attached to walls, trees, posts and poles.

However, there are some caveats. In a shaded area, such as woodland, a solar panel may be ineffective, and if the area is used by the public, installing a panel can attract unwanted attention.

Fuel cells

If finance is not an issue, the power problem can be solved with the use of fuel cells. These are used in caravans and boats to keep the battery charged, with the result that long periods can elapse before the battery needs mains charging. This is of particular use in very remote areas where regular visits to replace batteries are not feasible, and fuel cells are used, for example, by some television wildlife programmes. However, cost is a major issue: fuel cells cost several thousands of pounds.

Battery research

Battery research is intense due to the number of electronic devices requiring long-lasting batteries. For example, news has recently emerged of successful trials with lithium–sulphur batteries. If they live up to the claim, they will supply five times the power at a fifth of the cost, making portable power sources light, easy and cheap to use. This would make portable CCTV very attractive and flexible.

5.2.2.4 Image transmission components

Video connectors

To set up a portable CCTV system with good-quality images, it is necessary to ensure the best connections. Poor-quality connections can result in image degradation. Good-quality cable (RG59) with BNC connectors, or Cat5 cable, is essential for the best images. Most surveillance cameras have a female BNC connection, so any extension lead should have male BNC connectors.

Power connectors.

The power lead connections between battery, recorder, camera and setup monitor are not as critical as no video signal is present. In most cases 5.5 mm (outside diameter) 2.1 mm (inside diameter) 12 V DC power barrel connectors are used. These connectors are generally rated for a current less than 2 A, which is suitable as portable systems should aim for a total current less than 1 A to reduce battery power requirements.

5.2.2.5 Waterproofing

Cables

Cables should also be described as waterproof. Those that specify that they can be buried have the greatest water protection and are the best to use. RG59 cables are also available in indoor quality, so care should be taken when purchasing cables.

Cat5 cable can also be used, but Cat6 cable has greater water resistance than Cat5. The twisted pair wires in a Cat6 cable are surrounded by a gel and enclosed in a plastic sheath that increases water resistance.

Waterproofing connectors

The connections to the camera can be waterproofed with several layers of waterproof tape. Insulating tape sticks well to the connectors, and several layers of tape will give a waterproof seal for several weeks. The tape sticks so well it is a good idea to try and put one layer on in one piece as otherwise it can be very difficult to remove the tape afterwards. Connection boxes are available, but they are intended for cables to be pushed through and the connectors added afterwards. Adding BNC connections in the field is best avoided by those who are not well practised.

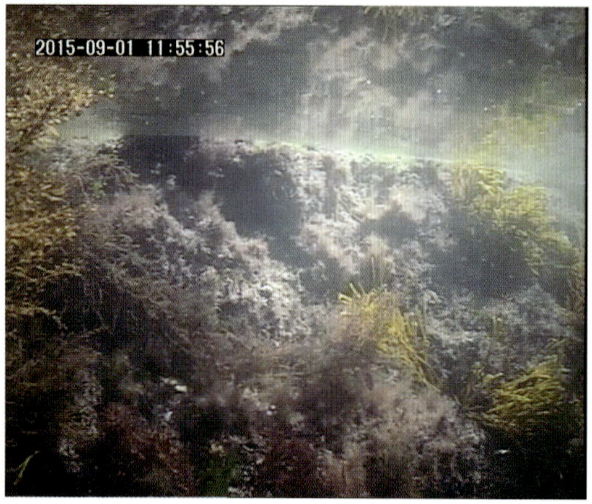

Figure 5.4 The fairy-tale world of the rock pool requires a submersible camera.

Waterproof cases

The cheapest method is to use a plastic box and seal any holes with sealant, but reliable waterproof protection can be achieved using high-specification waterproof cases such as Peli cases. They are easily drilled for connectors, but care must be taken that these holes are well sealed.

The connections to the case can also be made using Cat5 cable and baluns. A waterproof cable gland can be fixed to the waterproof case and the cable passed though. When the nut on the gland is tightened, this makes a waterproof seal. As power and CCTV baluns attach by screws, it is a relatively straightforward matter to connect them inside the case with the recorder.

Double-box system

Although a waterproof camera is essential for outdoor portable CCTV, the recorder, battery and setup monitor may not be waterproof. Waterproof equipment does exist (usually described as ruggedised) but for this one would have to pay thousands of pounds.

Total waterproof sealing can be a disadvantage if ventilation is required to prevent the DVR from overheating. I use a double-box system, which has several advantages. The inner box can be loosely closed or ventilation holes can be added to the sides, to allow air circulation. This may be required in summer, but not usually in winter as any heat from the DVR will soon be dissipated, but each system should be tested before any firm decisions are made.

An outer plastic box will compensate for any reduction in waterproofing, and will have space for surplus cable. It also reduces the need for high-specification waterproof connections to the inner case.

In cold, damp weather, condensation can form on the metal connections of the inner case. If this is a problem, they can be thermally insulated with non-conducting foam.

5.2.2.6 Security

Although clearly labelled CCTV cameras may deter theft because people assume they are being recorded, this will not be sufficient security for portable CCTV. The total cost of the components is higher and the prospective thief may realise that filming is not live. Extra precautions may be required.

In the field I wrap a chain round the outer waterproof box, pass it through the box handles then round a post or tree. The chain is then secured with a padlock. Camouflage material is placed over the box, then bracken, twigs and small branches from the site are placed on top. Any surplus cable is hidden under the camouflage. A camouflaged system is shown in Figure 5.5.

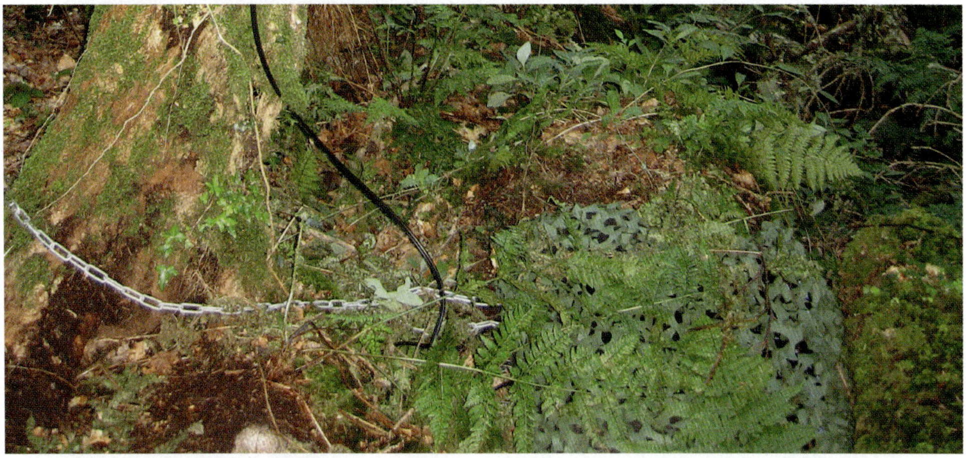

Figure 5.5 A camouflaged portable CCTV system. The boxes containing recorder and battery are on the right. The chain and camera cable are not yet camouflaged.

5.2.2.7 *Applications*

There are a wide range of applications that require portable CCTV and cover almost any form of wildlife monitoring where there is no mains power. Chapter 8 has examples which describe the use of portable CCTV, and Chapter 9 describes two examples of portable CCTV systems being used for fish and bats.

5.2.3 Portable CCTV setup

A detailed step-by-step guide to setting up a portable CCTV system is described in section 7.2.

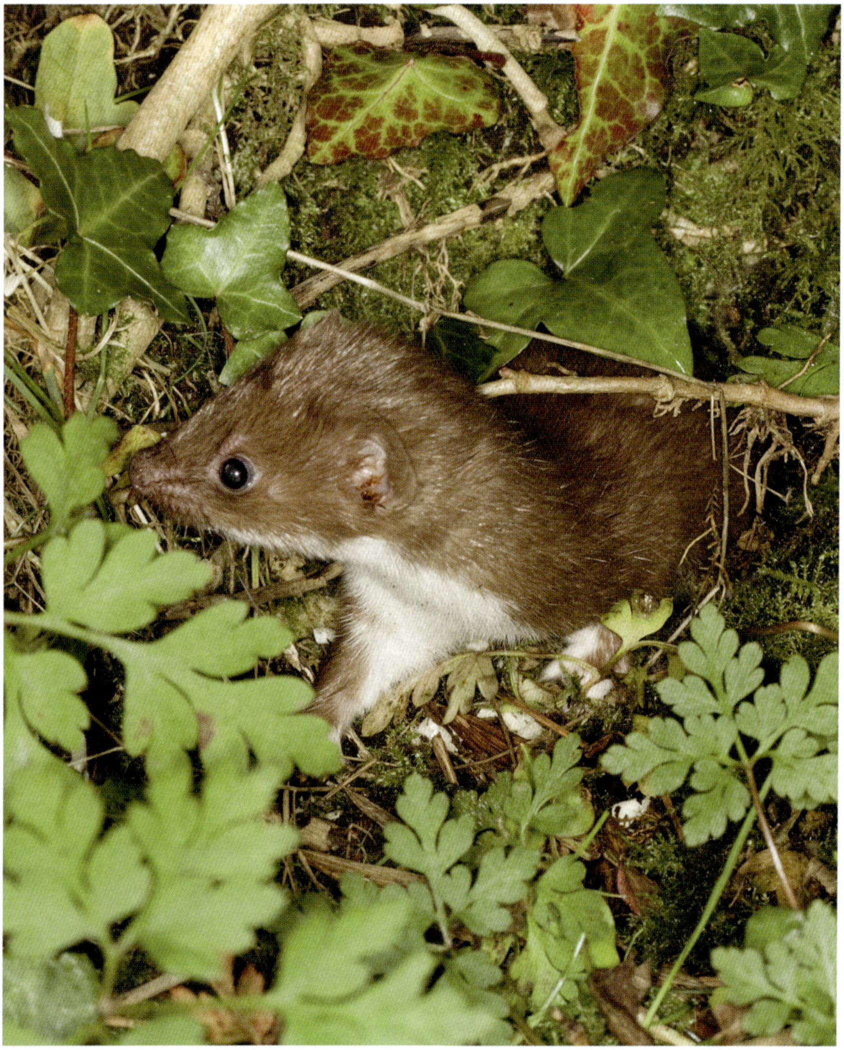

Figure 5.6 The fast-moving and elusive weasel (*Mustela nivalis*) is a good subject for portable CCTV. They are often seen near wood piles looking for voles.

6. Videos

There is a variety of methods for recording and processing videos. The standard recording device is a digital video recorder (DVR). These usually have the ability to record from 4, 8, 16 or more cameras, although a few have a single camera input only, which is often more suitable for wildlife monitoring. If cameras are to be connected to a PC, either via capture devices or, in the case of IP cameras, via a POE switch, software on the PC can be used to organise the recording and playback.

6.1 Digital video recorders

A standard CCTV recorder is shown in Figure 6.1. An HD-TVI recorder (see Figure 3.3) is similar in construction.

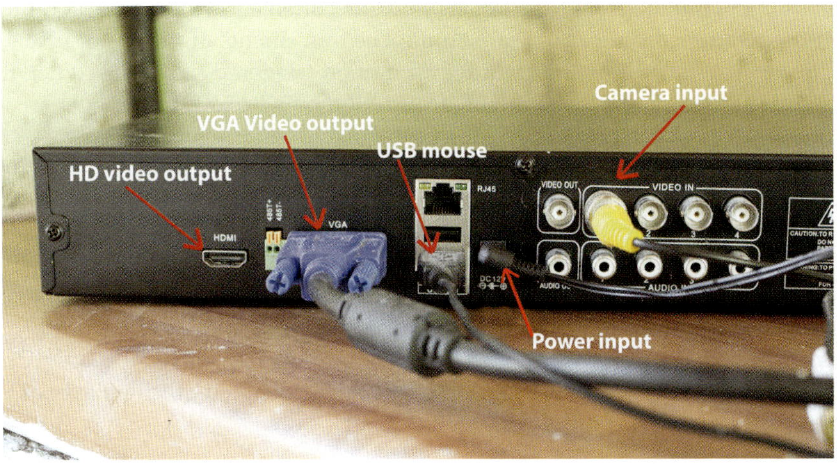

Figure 6.1 Digital video recorder.

6.1.1 Construction

The DVR is the standard method of recording videos, and, with the exception of portable recorders, is usually at least 35 × 25 × 5 cm in size. DVRs are in the form of a metal box with connections for a minimum of four cameras, but are also made with a greater number of camera inputs.

DVRs vary in construction but usually have female BNC input sockets, RCA input audio sockets, and video and audio output sockets. The recorder in Figure 6.1 has both

BNC, HDMI and VGA video outputs, as well as a USB mouse for using the inbuilt menu. Power in this case is 12 V DC, so that theoretically this recorder could be used with a 12 V DC battery, although the large current drawn would discharge the battery quickly.

Some DVRs have an embedded web server which means they can output video to the internet for remote viewing. As a result, analogue and HD cameras can be viewed remotely despite not being IP cameras. This feature is not straightforward to set up and a video-streaming service is the best option, provided that the recorder is compatible with the streaming service.

Most recorders require a hard disc and although a large number of videos can be stored, it is good practice to sort through the videos each day, selecting and exporting those required and deleting the rest.

6.1.2 Menus

DVRs usually have a menu that can be viewed on a monitor. The features vary, but usually allow adjustment to motion detection parameters, motion detection sensitivity, pre-record and post-record time lengths, time scheduling and other features. Some manuals are well written and very easy to understand, but others are not. Often it is possible to download the manual before buying, and reading the manual will give a good indication of how easy the instructions are to follow.

6.1.3 Recording

The recording parameters of DVRs are very flexible, often allowing time scheduling and motion-triggered recording at the same time. The movements of groups of pixels in the video image are analysed by the recorder for motion-triggered recording. When motion detection is set to high sensitivity, small groups of moving pixels will trigger recording, and for low sensitivity, larger groups of moving pixels will trigger recording.

Pre-recording is a useful feature whereby the recorder keeps 10 seconds or so of recording in a buffer, so that when motion triggers recording, the 10 seconds before the triggering took place are added to the front of the video. This means that there is no action missed from a slow video reaction time.

Masking is also possible so that a section of moving foliage, for example, can be masked so that although it will be seen in the video, its movement will not trigger recording.

Figure 6.2 gives an example of how motion may be denoted during playback.

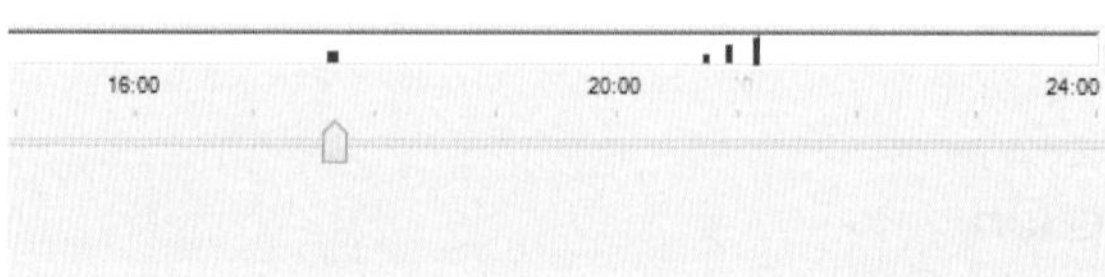

Figure 6.2 Section of a DVR playback timeline showing motion-triggered recording as red marks.

6.1.4 Playback

The playback system varies depending on the recorder, but when motion detection is used, the 24 hour timeline for each day's recording usually has coloured marks showing where motion took place. When playing, the recorder will jump to those marks. This feature, together with fast forward, makes it easy to scan a day's videos and note those portions of interest for export.

6.1.5 Exporting

Once the videos have been reviewed and the sections of interest have been noted, the export of videos for those time periods can take place. Usually recorders divide videos into smaller time periods, and the appropriate time periods can be selected for export. Videos may then need further editing using a video-editing package.

6.2 Portable DVRs

Portable (mobile) recorders are less common and very variable in quality and features. Mobile recorders are common for vehicles, but tend to draw a current of greater than 500 mA and often lack the motion-triggered recording and pre-record features required for wildlife monitoring. It is possible to get 'ruggedised' mobile DVRs, which are designed for use outdoors or in hostile environments. They are expensive but for long-term projects may be a worthwhile investment.

The recorder I use a great deal is a Genie SD-DVR, which is suitable for use with one camera. It is small and light and draws a small current (150 mA), making it ideal for portable CCTV systems. In addition, the recording parameters allow for scheduled and motion detection recording at the same time, together with pre-record, all of which are essential for wildlife monitoring.

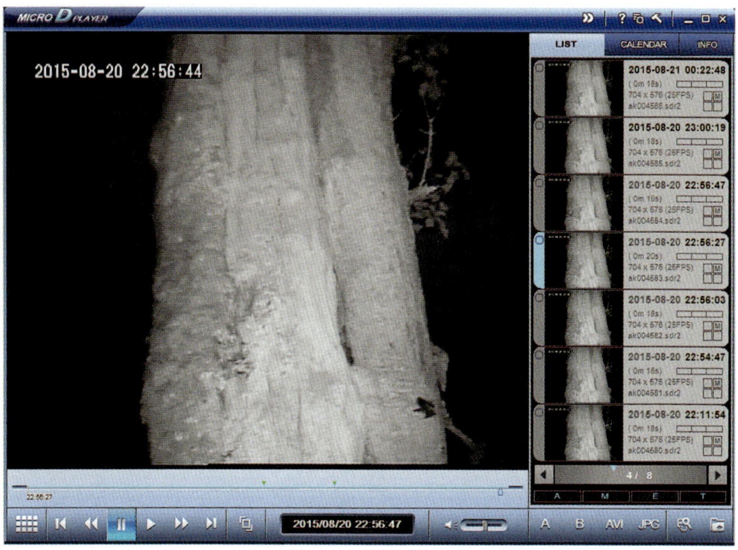

Figure 6.3 The Genie DVR comes with basic editing software.

This recorder is not designed for external use in damp or dusty environments, but my double-box system, described in Chapters 7 and 8, works well for me, giving a waterproof system that allows for ventilation when required.

6.2.1 Genie SD-DVR software

The Genie SD-DVR comes with its own software which makes it easy to view and edit videos before exporting them as AVI files. The column to the right in Figure 6.3 shows the video length and its start time, with each video having been time stamped. The length of time after a motion detection trigger can be kept short so that the videos can be quickly reviewed. If motion lasts longer than this time, filming will continue, as motion detection recording is triggered continually until motion stops.

The videos can be fast forwarded by ×16, and videos quickly reviewed.

To rough-cut videos, the button marked 'A' is pressed at the point you wish to start the video, and button B pressed when you want to end it. Pressing the AVI button exports the cut video as an AVI file which can be played on standard platforms such as Windows Media.

6.2.2 iCatcher

iCatcher software is designed by iCode. iCatcher comes into its own when used with IP cameras. It is sophisticated software, and not perhaps for the novice, but it has many useful features. It can be used with analogue cameras, but requires the addition of a capture card to the PC, unless only one camera is used in which case an external USB capture device can be used.

The playback screen in Figure 6.4 shows in red where motion has occurred. Videos can be exported by selecting the times when the video output is to start and stop.

iCode also host a high-quality streaming service.

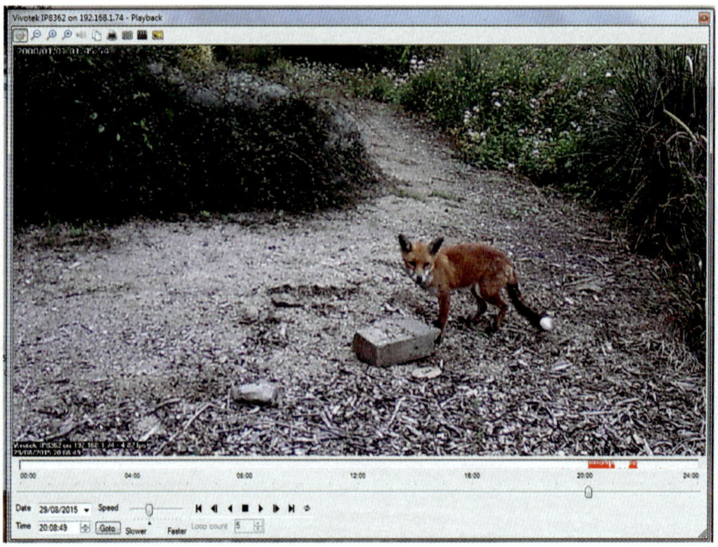

Figure 6.4 The iCatcher playback screen shows motion-triggered recording as red lines.

6.2.3 Video-editing software

You may be happy with the videos you have, but the ability to edit videos is a great advantage, whether this is because you want to show them on YouTube, or for more serious research. There is a great range of video-editing software which will allow you to add captions, titles, copyrighting text, or merge videos together. Top of the range is probably Adobe Premier Pro, but the internet lists many cheaper packages.

6.3 Dealing with videos

CCTV for wildlife monitoring is intended to be a pleasurable tool, whether for the wildlife enthusiast or for assisting with research. As such it is important to manage the tool well and not end up with the task of ploughing through thousands of long videos. You need to know and understand CCTV so that it works for you and not the other way round.

There are many factors to consider if you want to keep video numbers to a manageable size, and reduce the number of videos with nothing useful.

6.3.1 Project design

Project design is an essential part of any project, especially in scientific research. If you intend to use CCTV as a monitoring device, consideration should be given as to the likely number of videos that will be produced and how they will be processed. If it seems likely that a large number of videos will be the result and the resources to process them unlikely to be available, then that part of the project should be redesigned and particular attention given to the factors that can affect video numbers.

6.3.2 Reducing video numbers

The most obvious way to reduce video numbers is to use motion detection.

6.3.2.1 Motion detection

Setting up the correct motion-detection parameter for your target species is important. Recorders base their motion-detection recording on moving groups of pixels. Sensitivity can usually be changed and should be high for small, fast-moving creatures, and low for large creatures. Normal or average is for wildlife somewhere in between these two. High sensitivity triggers recording on small groups of moving pixels in the image, whereas low sensitivity uses larger pixel groups.

6.3.2.2 Wired or wireless

I cannot recommend the general use of 2.4 GHz wireless cameras for wildlife monitoring using motion-detection recording, and would even suggest they are unsuitable for use in urban areas. This is because they are frequent victims of interference and electrical noise, which not only affects video quality, but will also set off motion-detection recording.

However, there are situations where wireless systems are suitable. Monitoring seabirds on remote cliffs is one example where wireless CCTV can be an advantage and there is likely to be less interference. In such cases regular visits to change batteries and collect recordings would be hazardous. The camera could be put in place in one visit, with the camera power being supplied by a lead-acid battery kept charged by a solar panel. The video signal would be transmitted wirelessly to the recorder which could be situated on a more easily accessible part of the cliff. This would allow regular non-hazardous visits to collect data.

6.3.2.3 Post-recording length

This should be set as low as possible. Fast-moving creatures will either be gone quickly, or will continually trigger motion so that the final video length will be determined by how long the creature is in view. For example, with bats, I set post-recording to between 10 and 20 seconds. A falling leaf triggering the camera will only give rise to a very short video that is easy to view and discard, whereas a bat that moves for some time will keep triggering motion detection recording, so that the video will be longer.

For large, slow-moving creatures, post-recording should be set for a longer period, depending on the target species. Badgers, for example, can move around slowly if foraging, and the badger will only trigger motion recording now and then, so a post-recording time that is too short may not record all that it should.

6.3.2.4 Scheduling

As well as motion-detection recording, time scheduling should be used when possible. For nocturnal wildlife, the recorder should be set to record with motion detection from dusk to dawn. For underwater creatures in rivers, the limitations of light mean that recording should take place during daylight hours.

6.3.2.5 Background selection

When setting up a CCTV camera, choice of position can have a major bearing on video numbers. The camera should not have moving foliage in its view as that can trigger unwanted videos. Either change the camera position or indulge in a little light pruning.

Moving patches of light are another trigger, so setting the camera up in medium shade can be beneficial. Reflections from surface, especially water, can be another trigger. If this is a major factor, reducing the recording time can help. For example, if monitoring otters, scheduling recording between dusk and dawn will greatly reduce the reflection-triggered recording. Although otters can be around during the day, peak activity is at night, so you would have to weigh up the advantage of fewer unwanted light-triggered videos against the disadvantage of missing a daytime otter.

6.3.2.6 Masking

Many recorders allow masking as shown in Figure 6.5. This means that areas of the video scene can be masked so that, although still visible in the video, motion in that area will not trigger recording. Moving foliage or water are often areas that require masking. Care should be taken when drawing a mask as some recorders ask you to select the area to mask; others ask you to select the area to be viewed.

Software such as iCatcher also allows masking, shown in the red hatched area in Figure 6.5. The hatching can be made invisible once it has been positioned on the screen.

Masking is not always available with portable DVRs.

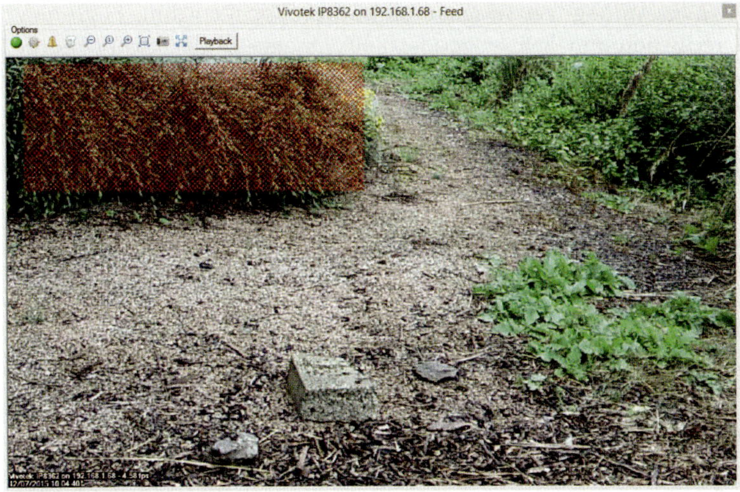

Figure 6.5 Areas that may trigger motion detection recording can be masked, as shown by the red hatching.

6.3.2.7 Weather

Rain and snow are a CCTV nightmare. Heavy rain or snow will trigger motion-detection recording almost continually and will also obscure the video. If such conditions are likely to continue for a period of time, it may be better to suspend filming, although the occasional day is not too much of a problem.

6.3.2.8 File size

It is easier to sort through several short videos than one long one, so file size should be set accordingly, if the recorder allows this. There is no perfect file size as it will depend on your preferences and the amount of data supplied by the camera. A high-definition video will be much shorter in time than a video from a low-resolution camera with the same file size.

6.3.2.9 Continuous filming

This is best used for occasional viewing as a large number of videos are produced. Wildlife centres often have video monitors with continuous viewing. This means that visitors can see what is happening at different times of the day. Streaming video is another example of a situation where the video is viewed live for short periods by individuals.

6.4 Picture quality

The picture quality of a video depends on many factors and control of these is vital.

6.4.1 Good-quality cameras

A good-quality camera will give a sharp picture with a stable image. Low noise is important as a noisy image can trigger motion-detection recording. High-quality cameras have noise reduction and deliver good-quality images under ideal conditions. A good supplier will advise you on this.

6.4.2 Good-quality cables

Outdoor-quality waterproof RG59 cables or Cat5 or Cat6 cables are necessary in order to reduce interference noise.

6.4.3 Image stability

If the video has lines across it, or if the image jumps every so often, this can trigger unwanted motion-detection recording and also give an unsatisfactory video. Cheap cables and poor connections are the main cause of unstable images.

6.5 Analysing videos

6.5.1 Exporting data

Recorders vary in their method of exporting data. Some recorders record to SD cards that can be read directly by standard media packages such as Windows Media player. Others require you to select the required videos and export them to a USB card. Some video recorders, such as the Genie SD-DVR, have their own PC software so that videos can be viewed, edited and exported.

There is no fixed standard of video file type as that depends on the compression parameters or codec used. Video-translating software may be required. There are many such packages available, although they are not all comprehensive. I use iSkySoft Video Converter, as it seems to cope with the most up-to-date formats.

6.5.2 Data storage

Mobile recorders usually record to an SD card, although a few record to a compact flash card. SD cards are not the most robust of data storage devices and if used roughly they may develop faults. It can be frustrating if, after several days of hard work lugging equipment miles and setting it up, the SD card has developed a fault and has not recorded. Careful handling is essential, in particular care should be taken not to touch the metal contacts on the back of the card. Cards should be kept in a card holder when not in use. If these simple precautions are adhered to, SD cards can be reliable.

The larger internal recorders usually record to a hard disc. Video data should be retrieved from the recorder at regular intervals and backups stored separately.

6.5.3 Analysis frequency

It is advisable to analyse videos at regular short intervals rather than leave it to the end of a project. Not only will this inform you so that you can adapt parameters, it also makes video analysis exciting rather than tedious as there will be fewer videos to analyse at a time. It is especially important for portable systems, and will alert you to the problem of a faulty card that has not recorded, or loose connections that have interfered with the video signal. Spiders can be a problem as they often spin webs over the lens, and frequent visits to remove them are advised.

6.6 Video enhancement

Video enhancement can make a big difference to a video that did not quite come up to expectations. This may be due to low contrast or a colour cast on infrared videos, low saturation of colour videos, or an image that is not quite sharp enough. Videos to be shown to others, for presentations, YouTube or research for example, often benefit from being shortened or having text added.

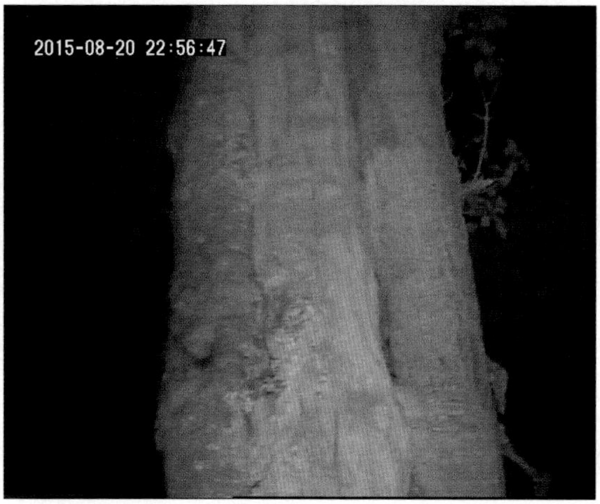

Figure 6.6 This video has low contrast and the camera is not quite focused correctly leaving the image looking soft.

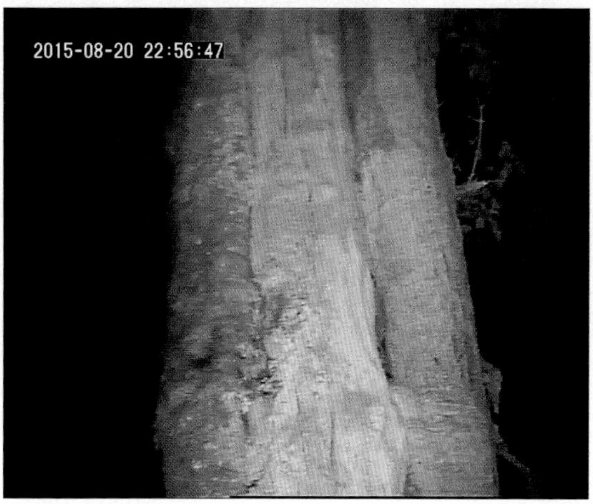

Figure 6.7 Increasing contrast and brightness, and sharpening the image, makes it clearer.

There are many video-editing programs, with Adobe Premiere Pro at the professional end and free packages on the internet for those short of funds. Any enhancements applied to a screen of the video will apply to the whole video, so at times a compromise may be required, as some scenes may need less adjustment than others. It is important to watch an enhanced video to the end to ensure some scenes are not over-enhanced.

6.6.1 Low contrast

This is a common problem with infrared videos. If a camera is set up for a suitable daytime contrast, it may not also be suitable for infrared videos. Changing the night video contrast may adversely affect daytime videos. The cheapest cameras tend to have the contrast fixed during manufacture, but as CCTV cameras have become more

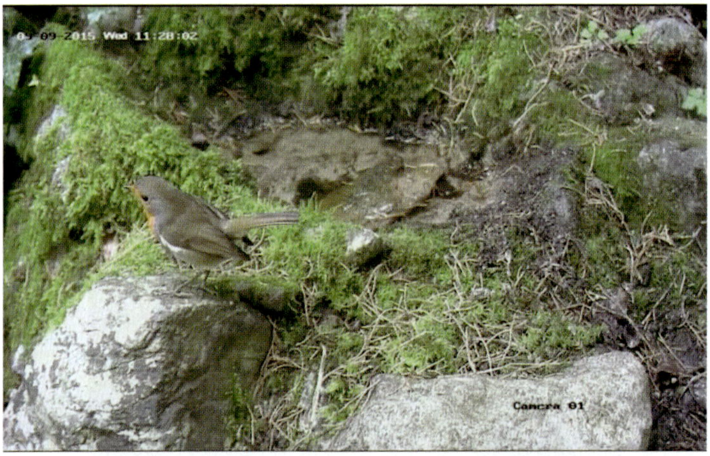

Figure 6.8 This video needs the right side trimmed and a few minor adjustments made.

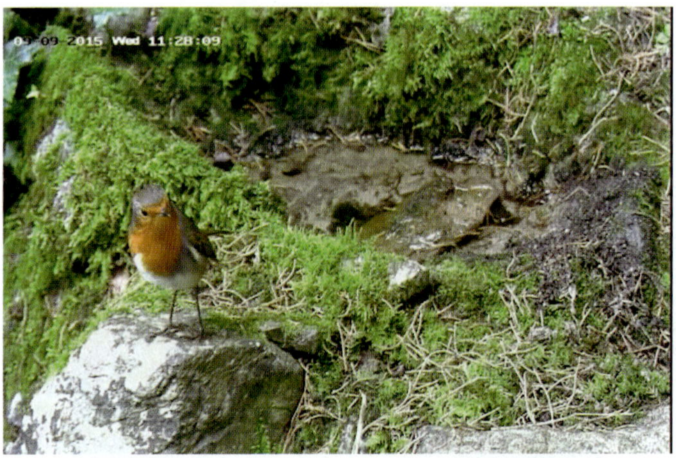

Figure 6.9 This shows the effect of a little cropping, followed by small adjustments in contrast, brightness and sharpening.

sophisticated, even low-priced cameras now have the facility to allow the alteration of contrast as well brightness, sharpness and other parameters. This is usually carried out by connecting the camera to a monitor and accessing the on-screen display menu (OSD).

The video in Figure 6.6, taken with a relatively cheap camera, demonstrates low contrast and a soft image. Using video enhancement (in Adobe Premier Pro), contrast and sharpness were improved (see Figure 6.7).

Even a good-quality video can benefit from some adjustments as shown in Figure 6.8 and Figure 6.9.

6.7 Videos

Low contrast and lack of sharpness are common problems. The camera in Figure 6.10 was clear and bright in visible light, but when the night LEDs switched on the image deteriorated. Adjusting contrast, brightness and sharpening can make all the difference.

The right side and the bottom of the video shown in Figure 6.11 were cropped to improving the framing of the curious robin. Small adjustments were made to contrast, brightness and sharpening.

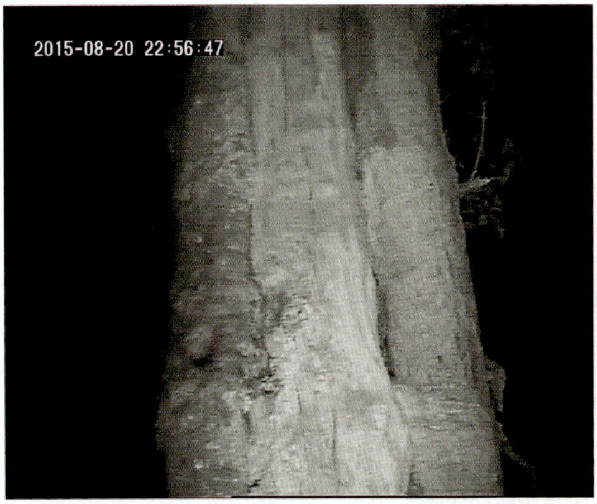

Figure 6.10 Enhancement to contrast.
https://youtu.be/TavyLIOZ9Ek

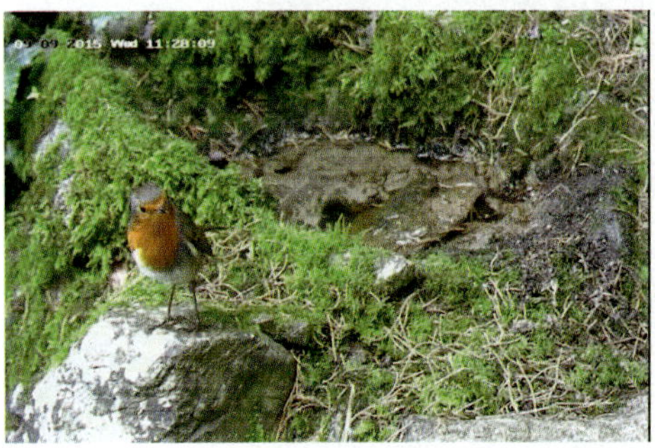

Figure 6.11 Cropped and brightened.
https://youtu.be/c1NGiHsa5lE

7. Technical case studies

The technical case studies in this chapter give step-by-step instructions on how to set up equipment for monitoring.

- 7.1 In pursuit of image quality.
- 7.2 Setting up portable CCTV.
- 7.3 Setting up an IP camera.
- 7.3 Making a lightweight mammal box.
- 7.5 Connecting a lead-acid battery.

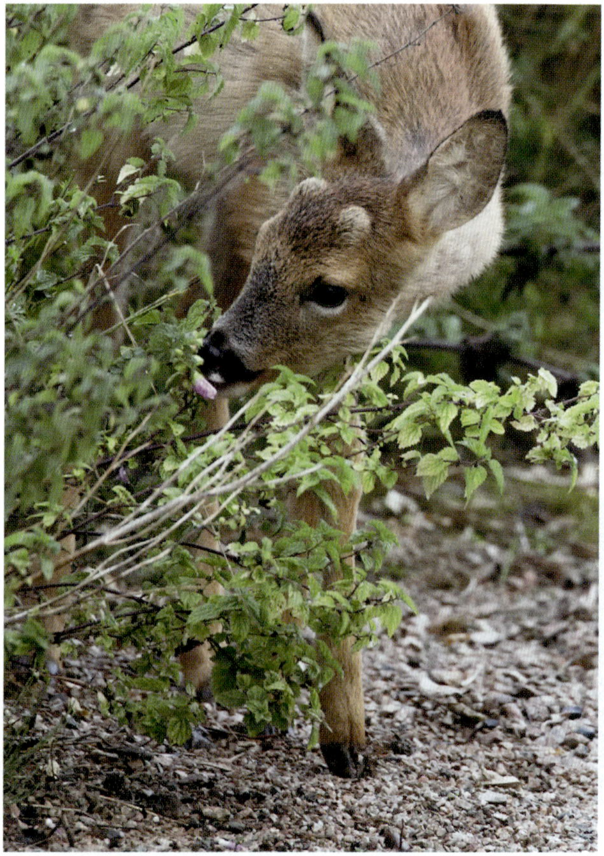

Figure 7.1 A young roe buck sniffs the flowers. Roe deer (*Capreolus capreolus*) make good CCTV subjects.

7.1 In pursuit of image quality

The novice's first CCTV videos may appear of low quality, especially when compared to those produced by broadcast-quality equipment (see Figure 7.2). However, it is possible to produce CCTV videos of quality with low-cost equipment, and this case study details a path to image quality that those new to CCTV might take.

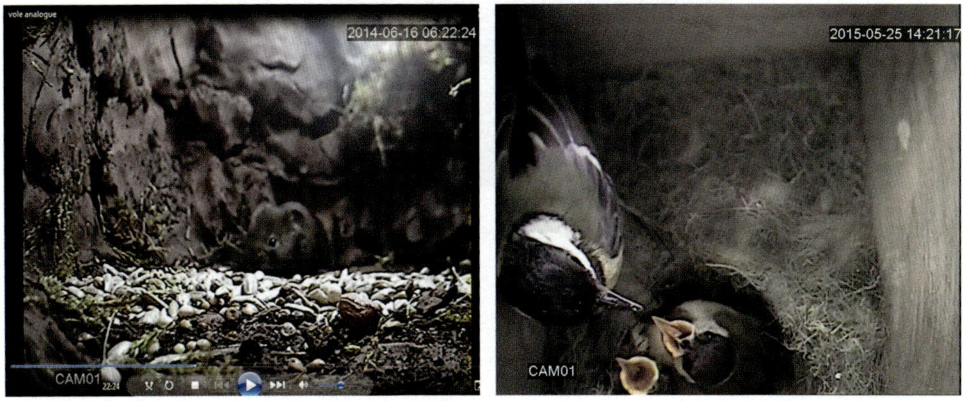

Figure 7.2 Images from a basic CCTV camera.

Step 1. Set up a simple bird/mammal box

The first CCTV videos are always a thrill, but after a while the quality of the images becomes an issue. Provided the images are kept small, they may seem adequate in clarity even with the small cameras found in bird boxes, but poor colour and lack of detail can become frustrating.

Small cameras tend to have basic features. For example, the colour may seem a little muddy. This may be the case with small cameras with infrared LEDs for night. The sensors on CCTV cameras are sensitive to infrared as well as visible light. The larger, more expensive cameras, termed as true day/night cameras or cameras with a cut filter, filter out infrared during the day, but remove the filter at night. For many bird box cameras, there may not be an infrared filter, with the result that colours are not true.

The cables used for bird box cameras are rarely RG59 cable and often are terminated with RCA connectors, with the result that images may be affected by noise or loss of image quality. The cameras with the highest resolution (700–1,000 TVL) can improve the image quality a little.

Step 2. Use an analogue bullet camera with RG59 cable

Bullet cameras come in many forms and there is a great deal of choice. The quality of the connecting cables and the connectors can have a great effect of image quality, as can the features of the camera. I have found that Genie and Twilight standard analogue cameras are of good quality, especially at 1,000 TVL. The suppliers recommended in the Appendix will help with camera choice.

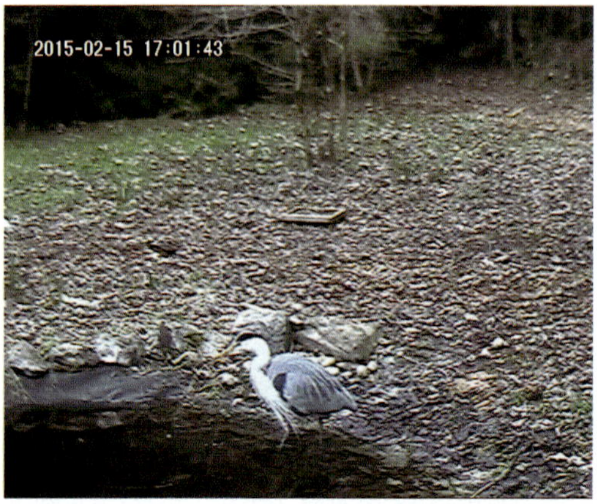

Figure 7.3 Poor connections can reduce image quality.

Step 3. Use good-quality connectors

The video from which Figure 7.3 was taken was of poor quality and the image jumped every 10 seconds or so. The reason was that although the cable was of good quality, the connections were not. Unless the video wire makes good contact with the connector, there will be a loss of image quality and may be an interruption of video streaming.

With good-quality cable and connectors, a video of the same scene in Figure 7.4 shows a higher-quality image.

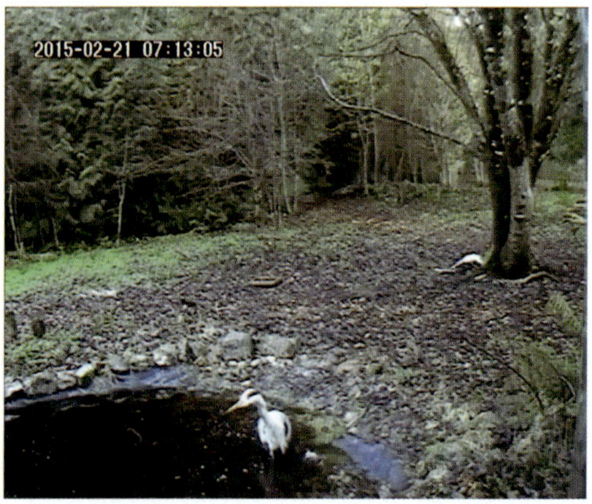

Figure 7.4 Improving connections improves image quality and stability.

Step 4. Increase camera resolution.

A typical image resolution for an analogue camera is 600 TVL, which equates to roughly 0.5 megapixels. Higher-resolution cameras are available, typically 1,000 TVL, which equates to roughly 1.3 megapixels. The difference is in the sensor and the sophisticated electronics in the camera which reduce noise, give even exposure and, with the use of a cut filter, true colour during the day and infrared at night. Fewer suppliers now sell lower-resolution cameras, due to the advent of HD analogue cameras.

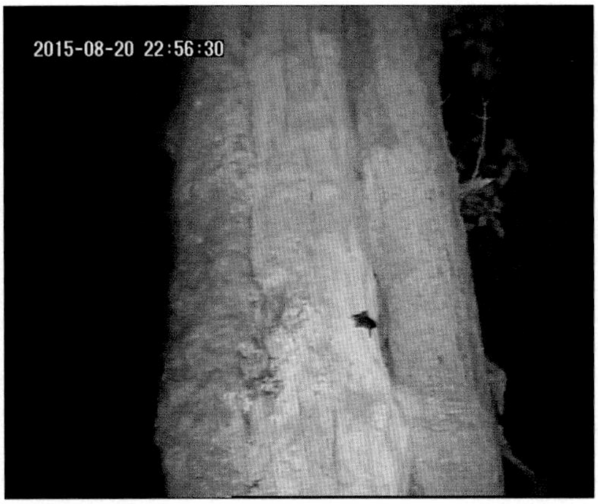

Figure 7.5 Image taken with an analogue camera with a resolution of 600 TVL.

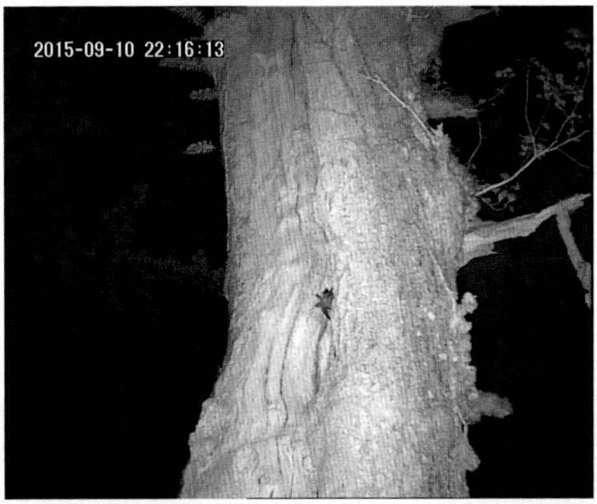

Figure 7.6 Image taken with an analogue camera with a resolution of 1,000 TVL.

Figure 7.5 was taken from a camera with a resolution of 600 TVL. The image is reasonably good and the bat shows up clearly. Figure 7.6 was taken from a camera with a resolution of 1,000 TVL and the higher-resolution image shows more detail and is enough to confirm that Figure 7.6 shows a long-eared bat (*Plecotus* sp.).

Step 5. Use an IP camera

IP cameras use ethernet cables to transmit their video signals and use the internet to show them on a computer. Even a small IP camera has a resolution of 1.3 megapixels, whereas larger cameras have resolutions of 2 megapixels upwards.

Colours are true, especially if the IP camera has a cut filter to remove infrared from sunlight during the day. The roe buck in Figure 7.7 was filmed in the early morning when light levels were low, but the image quality is good.

However, video quality comes at a price with IP cameras, as it depends on a good broadband speed. A slow speed will give jerky videos unless the frame size is reduced, and the advantage of using an IP camera may be lost.

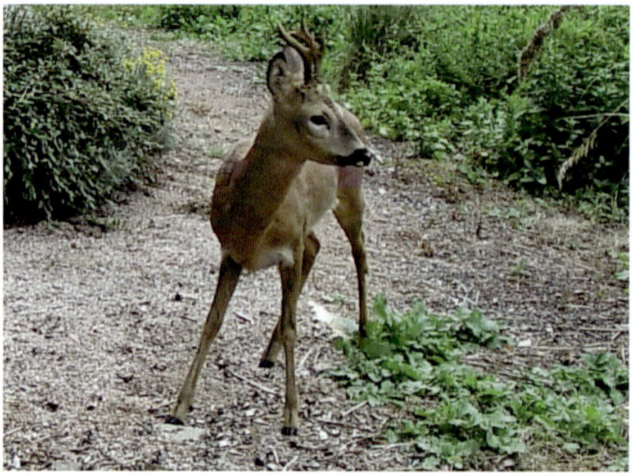

Figure 7.7 Roe buck filmed with an IP camera.

Step 6. Use HD-TVI

There are several forms of high-definition analogue systems, but HD-TVI not only gives HD images, it (at the time of writing) is much cheaper, and the indications are that it will overtake other HD forms of CCTV.

Colours are saturated and true to life with HD-TVI, and images have a great deal of detail, even at low light levels. As a result, stills taken from videos are of good quality as shown in Figure 7.8 and Figure 7.9. Because the signal is carried by normal RG59 cables, no internet connection is required. With HD-TVI the high-quality signals can be carried long distances, claimed to be 500 m before signal loss.

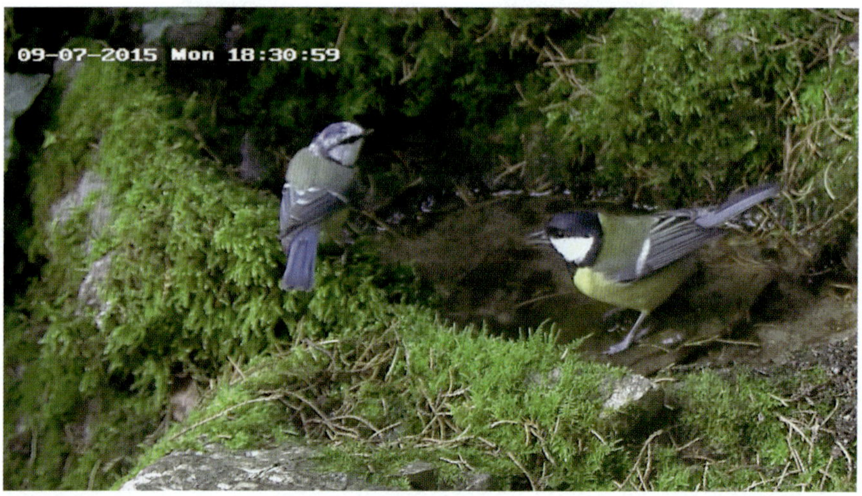

Figure 7.8 Birds filmed with an HD-TVI camera.

With HD-TVI, infrared night images are sharp and free from the grainy effect that is often seen with trail cameras. Figure 7.10 shows an example of HD-TVI night filming.

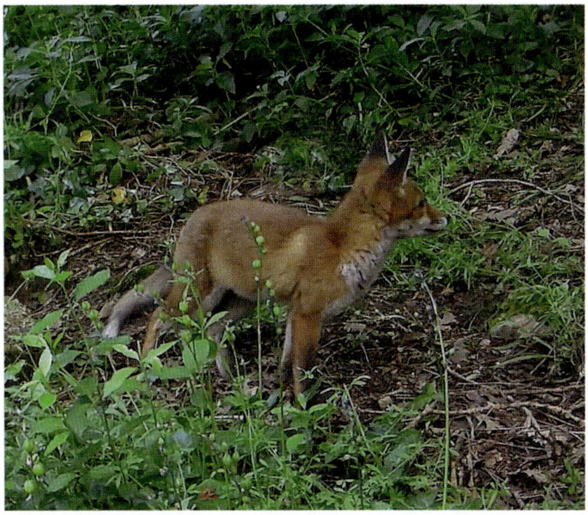

Figure 7.9 Fox cub filmed in HD-TVI.

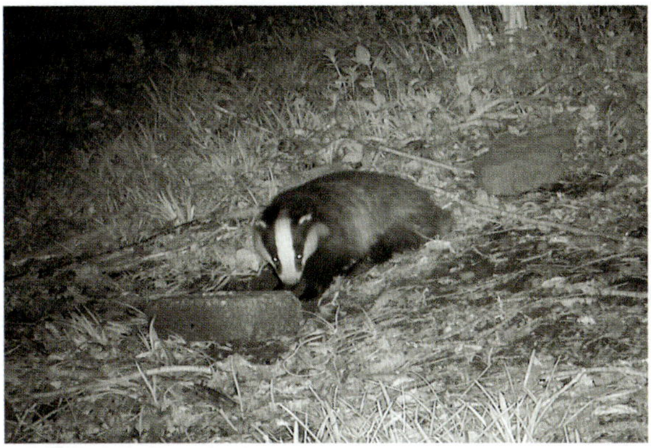

Figure 7.10 Badger filmed at night with an HD-TVI camera.

7.1.1 Conclusions

Considering the cameras available at the time of writing, the best-quality videos will be obtained most easily with good-quality cables, good connections and an analogue HD-TVI camera. Figure 7.8 and Figure 7.9 show the high quality possible with HD cameras.

7.2 Setting up portable CCTV

Portable (mobile) CCTV is the most flexible CCTV setup which, although restricted by the necessity for low-current devices, allows CCTV to be used in remote areas, or anywhere where there is no power, and especially in situations where a trail camera will not be suitable. Figure 7.11 shows a block diagram of a portable CCTV system (see also Figure 5.1).

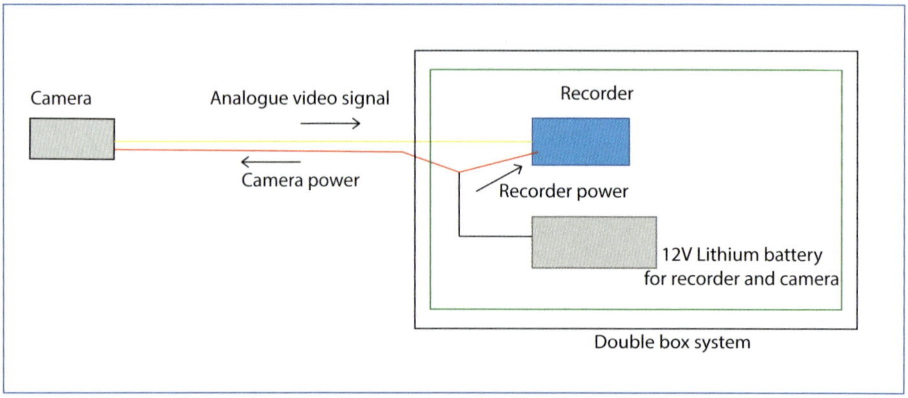

Figure 7.11 Block diagram of a portable CCTV system.

I designed this system to be truly portable, i.e. a system that could be carried by one person who had to walk several miles over rough ground. Battery, recorder, camera, test monitor and cables can be carried in a rucksack, with the lightweight boxes carried by hand.

Steps 1–3 should be carried out before arriving at the site, with the remaining steps carried out on site.

Usually placing the camera in position is a simple matter of screwing it to a tree, but for more complex arrangements such as using posts or climbing trees, another person, or an earlier visit, may be required to carry the additional equipment and set up the camera.

7.2.1 Equipment

- Peli case (or other sturdy plastic case).
- Bulkhead double BNC socket and cable gland. (Note: If you need to use fully waterproof fittings, you will also need to add waterproof fittings to the cable from the camera.)
- Large plastic box. Really Useful Boxes are ideal as they are strong, have locking handles and come in easy-to-hide black. It is easy to drill the necessary holes in the sides without them cracking.
- Analogue camera with low current draw (less than 500 mA).
- Mobile recorder such as a Genie CCTV SD-DVR. This recorder is not really intended for outdoor use, but I have found the double-box system seems to protect it. 'Ruggedised' mobile DVRs, suitable for outdoor use are available.
- Batteries. Lithium polymer batteries from Tracer are lightweight and convenient. They are not cheap but appear to be of high quality.
- RG59 Siamese cable to carry video signal and power.
- Power splitter with female socket and double male plug

7.2.2 Initial construction

Step 1. Drill a hole in the side of the Peli case and fit a bulkhead double BNC socket as shown in Figure 7.12. This is a socket which fits tightly against the outside wall and is tightened with a nut inside the case.

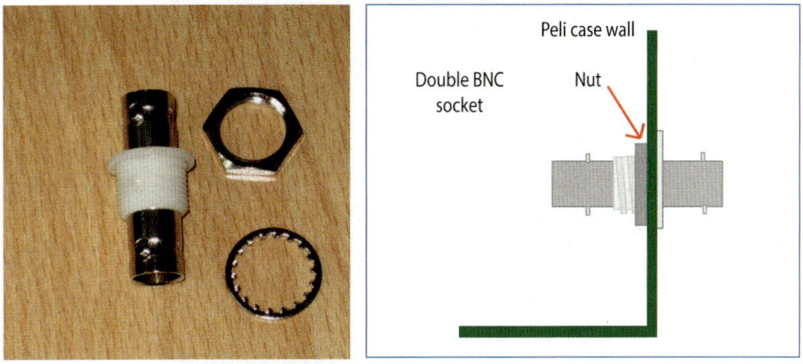

Figure 7.12 Double BNC socket for Peli case wall.

Step 2. Drill another hole to allow the power connector from the camera to fit through. If you want a waterproof connection, use a cable gland to hold the power connection. The power socket from the camera connects to the power plug from the battery inside the cable gland (see Figure 7.13). When the end of the cable gland is screwed in, it tightens the grip on the power socket from the camera, making a waterproof connection. A suitable size for the cable gland may be PG11, depending on the diameter of the power socket from the camera.

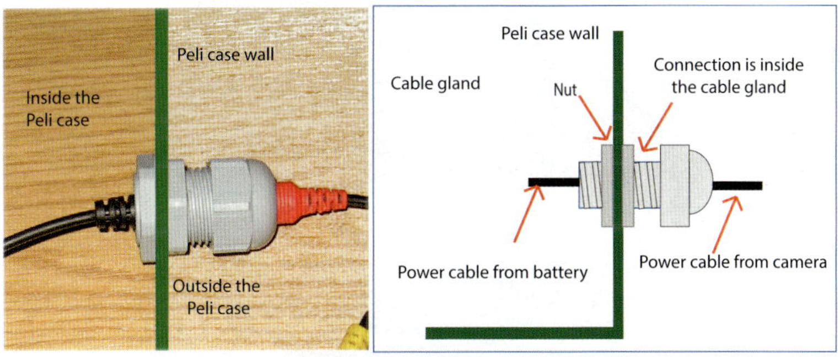

Figure 7.13 Cable gland for connecting power to the camera.

Step 3. Drill a hole in the side of the outer box to allow the cable from the camera to pass through.

7.2.3 Arriving at the site

Step 4. Position the outer box on stable ground and where it can easily be camouflaged.

Step 5. Open the inner box and connect the DVR 'AV in' connection to the inner BNC socket using a male BNC to male RCA adapter. Connect a male plug from the power splitter to the DVR power socket. Place the DVR inside the inner case. Make sure the SD card is firmly inserted into the DVR. The DVR is best placed on its long side to aid ventilation (see Figure 7.14).

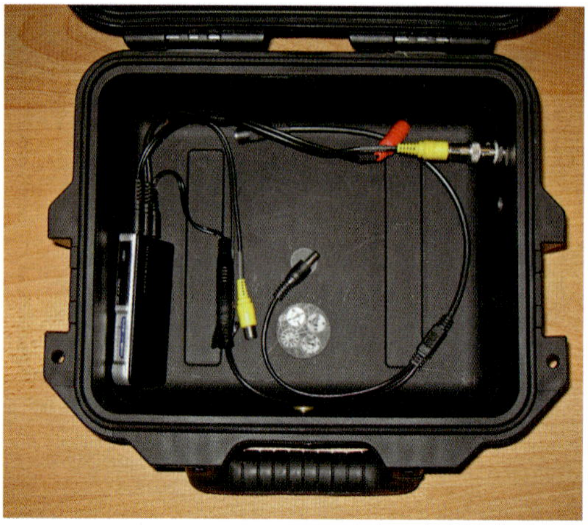

Figure 7.14 The DVR is connected to the video inner BNC socket and to one male plug on the power splitter. The red plug is not used. The unconnected yellow plug is for the test monitor.

Step 6. Fit the battery inside the inner case as shown in Figure 7.15.

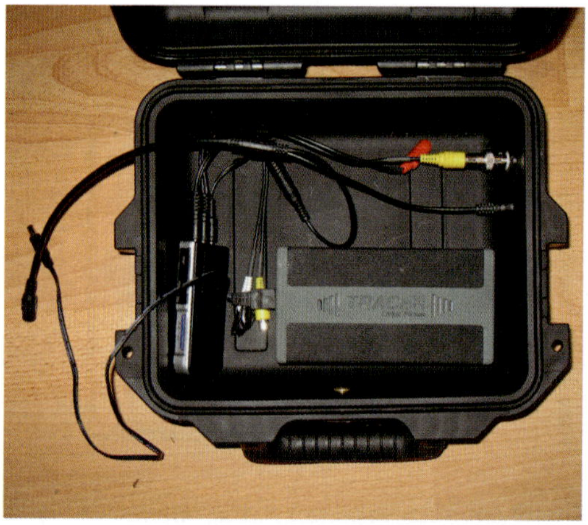

Figure 7.15 The battery is fitted into the inner case. The power splitter connection for camera power is pushed into the hole in the case. The battery connection is shown outside the case.

Step 7. Put the inner case inside the outer case. Pass the camera cable through the hole in the outer case. Connect the video socket from the camera cable to the outer BNC socket on the inner case. If a separate cable is required to lengthen the camera cable, ensure that the connections between camera and cable are made waterproof. This can be by use of a waterproof box or by wrapping several layers of adhesive waterproof tape round the connections.

Step 8. Connect the power socket from the camera cable to the free male plug from the power splitter through the hole in the inner case. For waterproof connections use a cable gland in the hole.

Step 9. Connect the battery to the female socket of the power splitter (shown outside the case in Figure 7.15).

Step 10. Connect a test monitor to the 'AV out' video socket from the DVR (the other yellow socket in Figure 7.15) to check that there is a signal and that the camera is positioned correctly. The unused white socket is for audio.

Step 11. Ensure that the recording settings on the DVR are set to your requirements using the DVR menu and the remote control for the DVR. The menu may not work if the DVR is recording. If this is a problem, disconnect the video cable from the camera to stop recording, use the menu, then reconnect the video cable when finished.

Step 12. Disconnect the test monitor and carefully lower the lid of the inner case, making sure no cables are caught. In warm weather do not close the lid tightly as this will restrict ventilation for the DVR unless you are using a sealed waterproof system, which has its own ventilation method.

Step 13. Close the outer box (Figure 7.16). Pass a chain under the box, through the box handles and round a tree or rock. Padlock the chain so that the outer box cannot be opened. Cover the outer box with a square of EPDM pond liner and camouflage net.

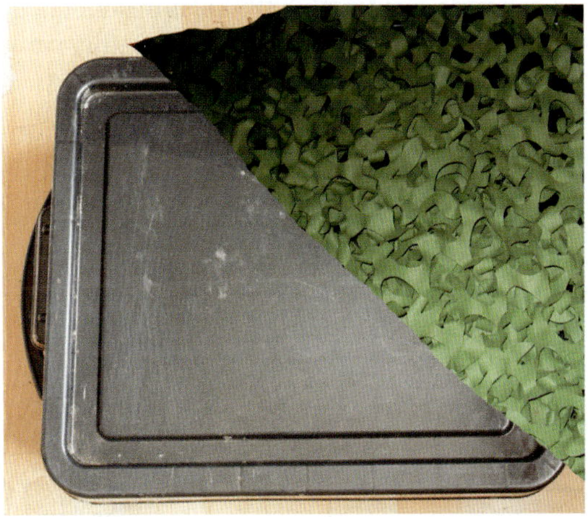

Figure 7.16 The outer box is closed. Suitable camouflage material is shown on the right. Natural materials from the site can also be used. The security chain is added before the camouflage.

When changing the battery and/or the SD card, make sure the DVR is not recording. Follow the instructions with the recorder to switch off recording, then disconnect the battery and remove the SD card.

I have found the double box system to be flexible and to work well for general wildlife monitoring. As the inner box is placed in an outer waterproof box, there is no need for fully waterproof fittings. I also prefer to keep the DVR and battery loose and not fixed in the case. This allows me to carry the DVR and battery in a rucksack, and also gives more flexibility with battery size and recorder type.

However, if you wish to use a portable system for inhospitable surroundings, rugged equipment in a sealed box with waterproof connections should be considered. In such a situation you may prefer to fix the DVR and battery inside the case.

7.3 Setting up an IP camera

Chapter 5 on IP cameras gives the outline for setting up an IP camera. However, the process is not straightforward, so a step-by-step approach is given here. The instructions are for a basic plug-and-play system.

7.3.1 Equipment

- IP camera.
- POE switch.
- PC.
- Ethernet cables.
- Mains power.
- Internet router (hub).

IP camera: There is a wide range of IP cameras, with Vivotek often being used for wildlife. For the purposes of this example, the Vivotek IP8152 is described, but the basic principles will be similar for other IP cameras. The advantage of this camera is that it is 'POE enabled', which means that it uses the ethernet cable to supply its power. This camera does not have integral LEDs so a light source may be required.

POE switch: This provides the power for the camera and receives the video signal from the camera. Connections are shown in Figure 7.17.

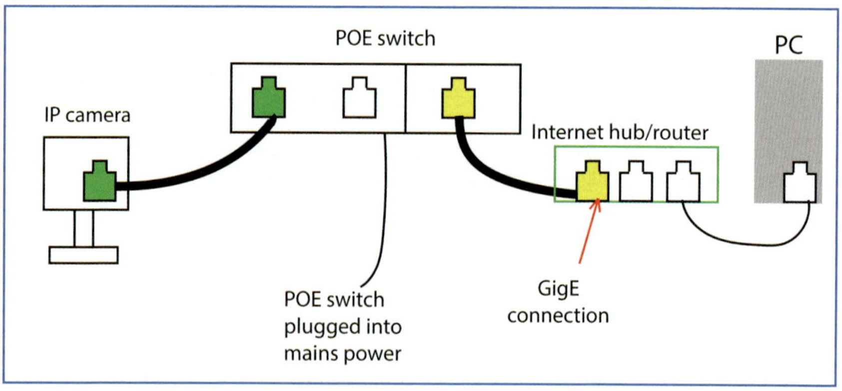

Figure 7.17 Connecting an IP camera with a POE switch.

7.3.2 Setting up

Step1. An ethernet cable should be connected to the camera using the socket on the camera. The other end should be connected to a socket on the POE switch input side (green sockets in Figure 7.17).

Step 2. Connect another ethernet cable from the POE switch output to the GigE input on your router (yellow sockets in Figure 7.17). This faster socket is more suitable for videos.

Step 3. Connect the PC to the router using an ethernet cable. Alternatively, WiFi can be used.

Step 4. Connect the POE switch to the mains power supply.

Step 5. Install and run the camera software.

For the Vivotek IP8152, the software will analyse your system and come up with a screen as shown in Figure 7.18.

Step 6. Clicking on 'Next' brings up the screen in Figure 7.19:

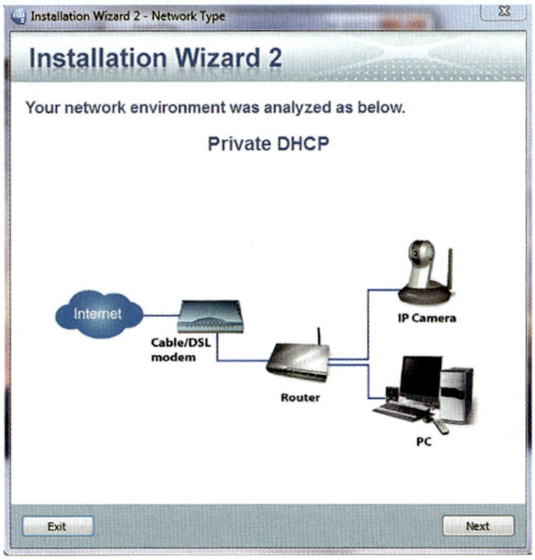

Figure 7.18 Starting the Vivotek camera software.

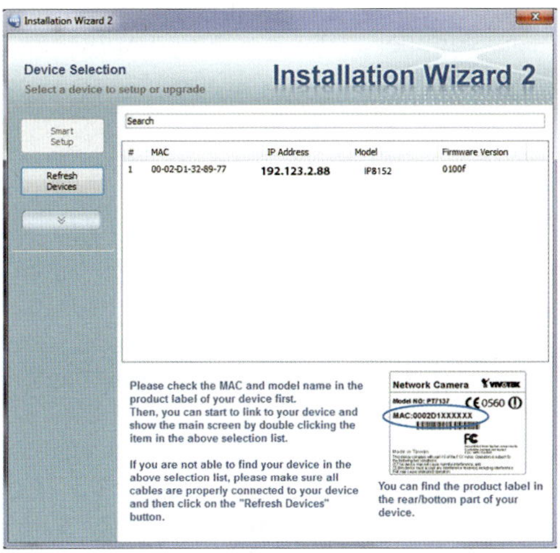

Figure 7.19 This screen gives the camera's IP address.

The camera has been assigned an IP address 192.123.2.88 (not a real address). This address is required if you wish to use recording software on your PC such as iCatcher. Each camera will have a different address.

Step 7. Double click on the item in the list. This will connect you to the camera so that you can view what the camera sees, and can adjust image parameters (see Figure 7.20).

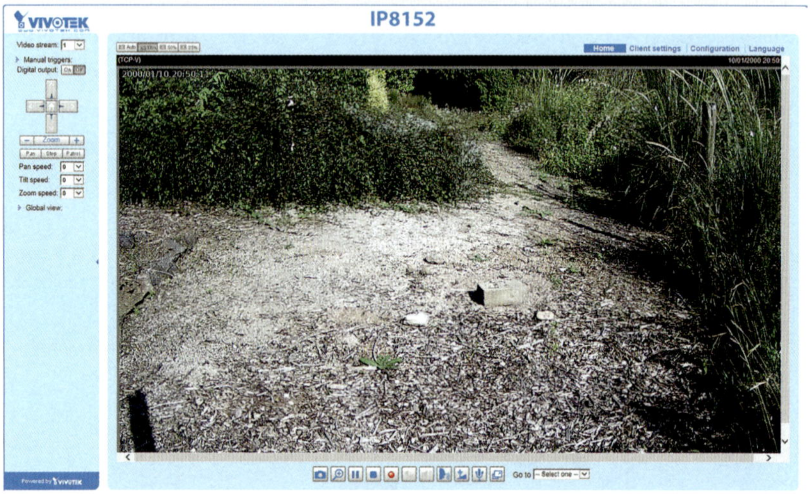

Figure 7.20 Once the camera view is visible, the menus can be used to adjust the video parameters.

Clicking on the menus along the top allows adjustment of parameters for image and video. Initially it may be easier to stick with default parameters. The resolution of the video, the size of the view and frames per second of the video are a few of the parameters that can be adjusted.

There are many other parameters that can be adjusted, with many applying to network characteristics. It is unlikely you would have to become involved with those, but if in doubt contact your supplier as this is a very complex area.

7.3.3 Output

7.3.3.1 Video size

The IP 8152 is a small camera ideal for mammal boxes. It can be used with a variable-focus lens which increases its flexibility. Its resolution is 1,280 × 1,024 pixels. It is possible to use IP cameras with a greater resolution, but this can cause problems with viewing the videos. The major difference between an IP camera and a non-IP camera is that the image is viewed directly via the internet. If you have a very fast connection (20 Mbps and higher) you will probably have no problems, but if your broadband speed is low, it will struggle to send the images fast enough. The result is that the resolution may have to be reduced, otherwise jerky or frozen videos may result, which rather defeats the purpose of using a high-resolution IP camera. Once again a friendly, knowledgeable supplier is essential.

However, if you only want to take stills from a high-resolution video, a jerky video will not be a problem.

7.3.3.2 Streaming live

Live streaming is one of the great advantages of IP CCTV cameras, although it can also be used with analogue cameras, provided the recorder has an embedded net server. Live streamed images can be seen on iPads and phones as well as PCs and tablets. This means that you can watch live even when many miles away, an ideal situation for a researcher or a group of people.

However, it is not as simple as it seems, as live streaming is not easy to set up and may hog all of your bandwidth, leaving nothing left for regular internet viewing. The answer is to use a streaming service, especially if more than one person is to view the camera output. The cost of this varies according to the size and resolution of your videos, and also with the streaming service itself.

7.3.3.3 Recording

Vivotek does provide software that can be downloaded for recording videos, but many use iCatcher, software produced by iCode for just this purpose.

7.4 Making a lightweight mammal box

The basic components of a lightweight mammal box are shown in Figure 7.21. The box can be made of any suitable material, but the easiest and cheapest method is to use two

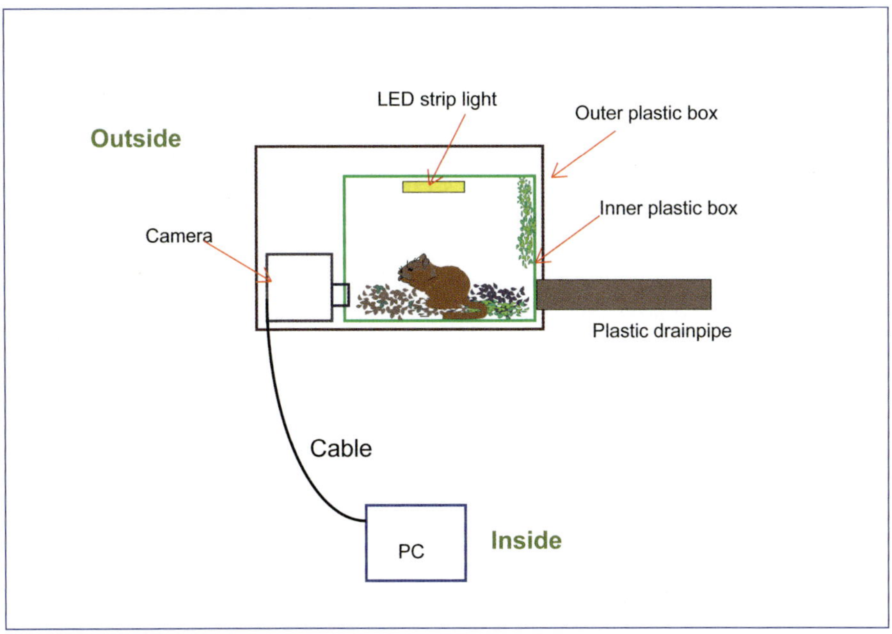

Figure 7.21 Mammal box set up.

plastic boxes. The smaller inner box can even be made from an old washing-up bowl with the rim removed. Plastic is easy to drill and can be washed, if necessary.

7.4.1 Steps to making the box

Step 1. Make or buy an inner box of roughly 20–30 cm square.

Step 2. Make or buy an outer box that will accommodate the inner box plus camera and stand.

Step 3. Drill a hole in the inner box so that the camera lens and infrared LEDs (if any) are just uncovered.

Step 4. Drill a hole at the other side of the inner box of the minimum size that your chosen mammal can enter. For example, 20–25 mm will allow vole entry and should deter all but the smallest rat. If there are problems with unwanted larger mammals, attach one of the metal plates used on bird boxes to restrict the diameter of the hole. This will also prevent rats and squirrels from chewing the hole to make it bigger. Step 4 is shown in Figure 7.22.

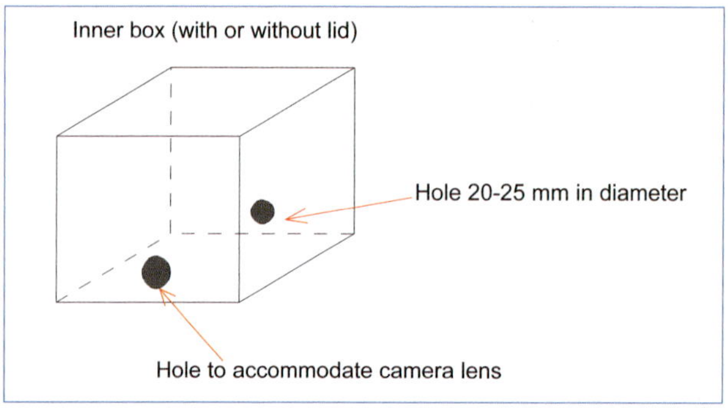

Figure 7.22 Mammal box step 4.

Step 5. Drill a larger hole in the outer box so that a plastic drainpipe can fit through. I used a drainpipe of 50 mm diameter as this allows mice to pass each other, and also gives them a dark area in which to eat. I did try a narrower tube initially, only to find one vole leaving the box from one end and another vole arriving from the other end. Neither would give way and some manual traffic control was required to prevent tube rage. Drill a smaller hole on the other side to allow cables to enter. Step 5 is shown in Figure 7.23.

Step 6. Ensure that the camera sits between the two boxes and is stable. It may be necessary to make a stand. I found that a wooden stand was required, as the supplied support was at the wrong height, but this will depend on the camera used (see Figure 7.24).

Step 7. The fun part is lining the inner box to give it a natural look. This is both for the benefit of any visiting mammals so that they feel at home, but also because it makes for a more natural background for videos and still images.

A wooden box can be lined with reinforced clay of the type used in schools. This is safe to use (both for you and the mammal visitors), but is expensive and very heavy, as about 4 kg is required to line a box. I have found it dries out too much in hot weather

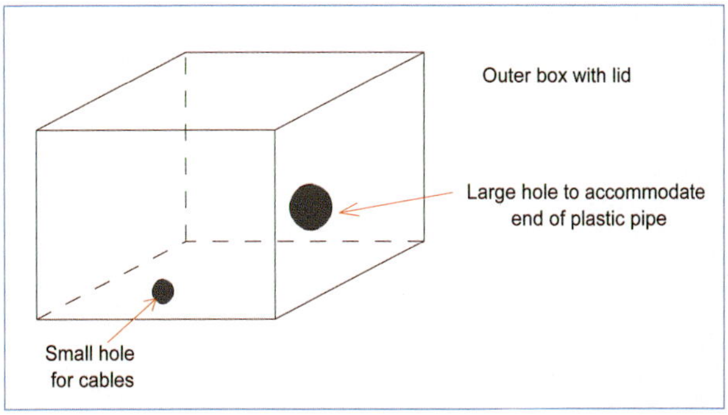

Figure 7.23 Mammal box step 5.

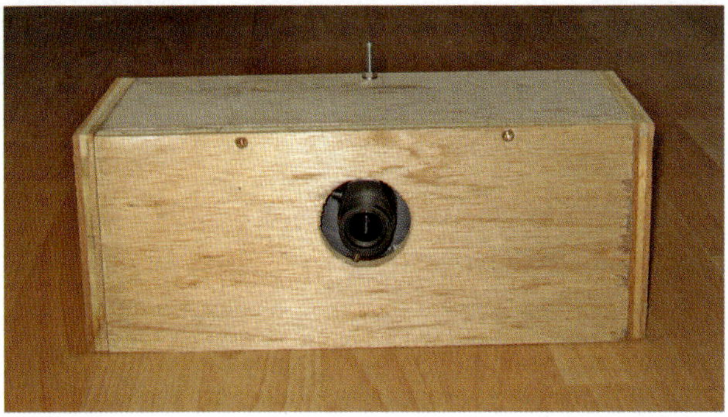

Figure 7.24 Camera stand.

leading to cracking, and in a very damp atmosphere can absorb too much moisture and become soft.

The best material I have found, especially for plastic boxes, is papier mâché powder. It is very light, inexpensive and pleasant (if sticky) to work with. A 1 kg bag is enough to line several boxes. The powder is mixed with water, kneaded until smooth then pressed onto the side of the box. Because it is very sticky, it adheres tightly to the box sides. While it is still damp, moss can be pressed into it, giving a natural look. The floor can be strewn with leaves and small twigs, rather like a woodland floor (see Figure 7.25). There is no need to cover the side that the camera is on as it will not be seen by the camera.

A lid is a good idea as it stops the mammals jumping out into the outer box (which they will certainly try to do) but it is not essential as long as they cannot get to the camera.

Once the papier mâché has dried, it can be painted, but I have found it easier to mix the dry papier mâché with children's poster paint powder before adding water. This is non-toxic (which paint may not be) and also gives the papier mâché a natural look, with brown or dark green being the best colours.

Step 8. When the papier mâché is dry and hard, the camera should be placed in position and tested to ensure that all areas seen by the camera are camouflaged. The two can be placed in the outer box, with the cable passing out through the outer box.

It is essential to have a camera that allows adjustment of field of view and focusing. Even with these controls there is always a trade-off between a wide field of view which shows the lid of the box, or a narrower field of view that shows only the inside of the box

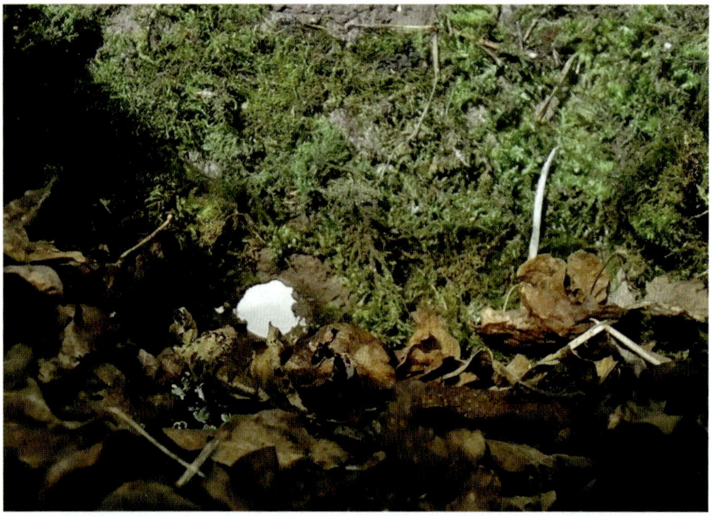

Figure 7.25 Internal view of a lined mammal box.

but which may miss any animal action to the side. One solution is to put moss on the part of the lid in view and choose an average field of view (see Figure 7.25).

Step 9. With a clear lid on the outer box, there may be enough light during the day to film in colour, but the images will be brighter with a light source. 12 V DC LED illuminators can be used, but 12 V DC LED strips are easier to use and very convenient. They are self-adhesive and can be stuck to the lid of the box. If you want to video visitors at night, an infrared illuminator can be used. Infrared LED strips can also be used, although these are hard to find in the UK.

Place the strips nearer the camera end of the lid so that they are not seen by the camera. With moss at the front of the lid, where the mammals enter, and LED strips nearer the back, it is possible to light up the scene and keep it looking natural.

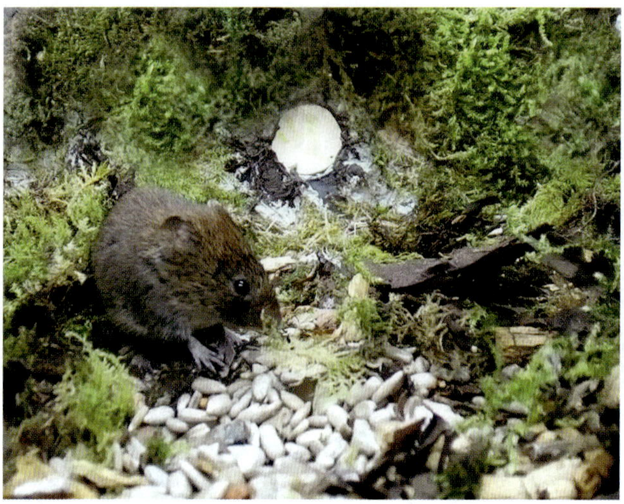

Figure 7.26 A mammal box visitor.

Step 10. The disadvantage of lighting is the need for a power source. If a standard CCTV camera is used, the power to the camera can be split so that the power to the camera is shared (if both use 12 V DC). If an IP camera is used, a separate power cable or a battery will be required to power the illuminators.

The 'decoration' inside the box will change quite quickly. The vole in Figure 7.26 covered up the seeds every night with leaves and moss, leaving the walls a little bare. It is an easy matter to repair any gaps.

7.5 Connecting a lead-acid battery

Lead-acid batteries are commonly used in vehicles whether cars, caravans, boats or golf trolleys. There are many connectors for these batteries, but they are not always suitable for applications where a lead-acid battery is used to supply power for a CCTV system. This section describes one way of making the connections. There are other connectors that would be just as suitable.

7.5.1 The battery

Sealed lead-acid batteries have a variety of terminals, and for illustration I have used a Numax leisure battery that is described as being suitable for electronic components. The battery terminals are a double-pole system, which gives more flexibility in choosing connectors. The smaller pole is for use with crocodile clips, so the instructions following refer to the larger pole or terminal post (see Figure 7.27).

The larger positive and negative poles are of different diameters. This particular battery has the following pole diameters:

- Positive: 19.5 mm, 13 mm.
- Negative: 17.9 mm, 13 mm.

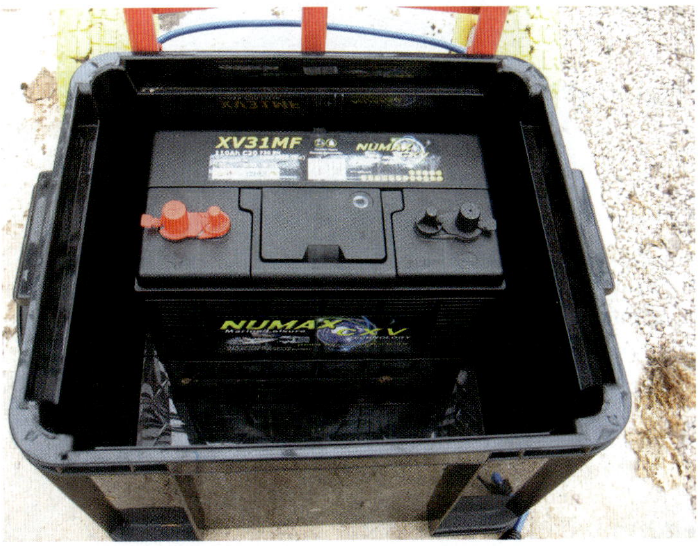

Figure 7.27 Lead-acid battery in its transportation box. Each of the double pole terminals has an insulating cover: red for positive, black for negative.

7.5.2 The connectors

A suitable connector is shown in Figure 7.28 and Figure 7.29. The end of the wire is attached to the screw on the connector. The wire may have bare ends which are wound round the screws, or they may be pre-fitted with an eyelet which fits over the screw. An eyelet is more secure and less likely to come loose. There are other ways in which the wire can be connected, but this is the easiest.

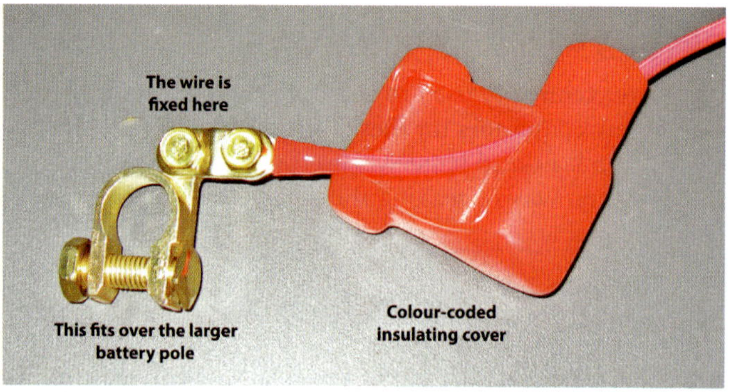

Figure 7.28 The connector for the positive pole.

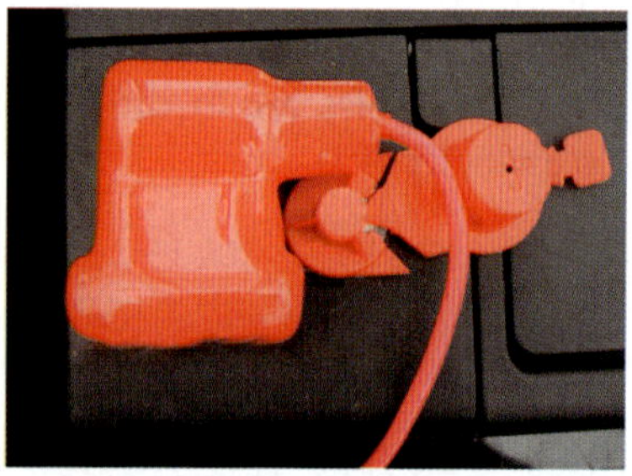

Figure 7.29 The connector is fitted over the large pole and the nut tightened. The plastic cover is pulled over the battery terminal connector. The smaller pole is covered with the cap supplied with the battery.

Red and black wires with a cross section of approximately 6 mm^2 are suitable. Thinner wire will not be held properly by the screws on the connector. The larger part of the connector fits over the larger pole of the battery, and the connector bolt is tightened.

It is advisable to use colour-coded plastic covers to fit over the connector. This gives some protection from water and makes shorting of the battery less likely.

A 1 or 2 A fuse must be placed in the live (positive) wire. It is easiest to buy wire with a fuse already incorporated to avoid having to solder one in place. These can be found in auto shops and on eBay.

7.5.3 The terminal block

The thick battery wire has to be connected to thinner wire so that suitable connections can be made to the camera and recorder. A covered terminal block is suitable.

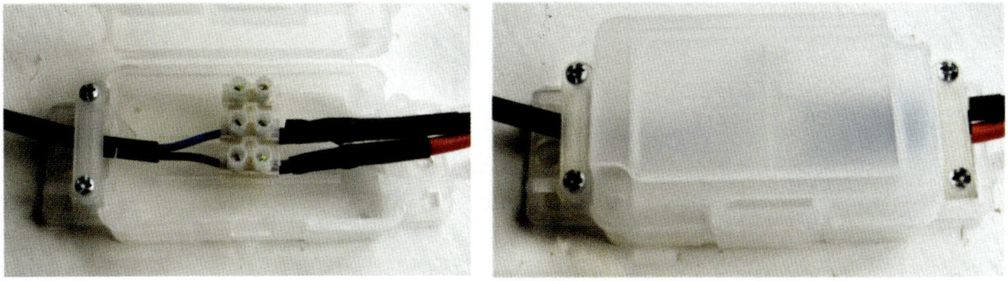

Figure 7.30 The wires are screwed into the terminal block (on the left) and the lid on the water-resistant box is closed.

The thick wires from the battery are pushed in the holes on the terminal block and screwed in place. Thinner wires from a power cable are connected to the other side. Positive must be connected to positive and negative to negative using colour-coded wires. Although the battery is being used to supply currents of less than 1 or 2 A, it is good practice to enclose the terminal block in a water-resistant box (see Figure 7.30).

The other end of the power cable can be connected to a power balun, in this case a barrel male connector with a 2.1 mm inner diameter. This will allow it to fit a power splitter and supply power to the camera and the recorder.

Great care must be taken at all times to ensure that the correct fuse is used and that the battery is not shorted by allowing positive and negative wires to touch. This can cause sparking and may damage the battery. It will also blow the fuse. If a suitable fuse is not used, the resulting large currents from shorting might even cause the battery to explode. For further details on battery safely see the Health and Safety Executive information on *Using storage batteries safely* (see References).

The battery should be transported without the connectors, which should be added when the camera and recorder are connected up.

7.6 Videos

Squirrel

The squirrel (*Sciurus carolinensis*) shown in Figure 7.31 is lost in contemplation at the beginning of this video before springing into action. This is an example of poor video quality caused by poor contacts and a poor-quality analogue camera. The image of the squirrel breaks up as it moves and the whole video jumps every few seconds.

Vole

The video in Figure 7.32 was taken in a mammal house and shows two voles (*Clethrionomys glareolus)* fighting, which mainly consists of chasing each other around. At one point one vole stands at the entrance, boxing the other to keep it out. The broadband managed to transmit the data smoothly as it was a small scene, but even 25 fps is not enough to cope with the speed of the chase, and the voles rush past in a blur at times.

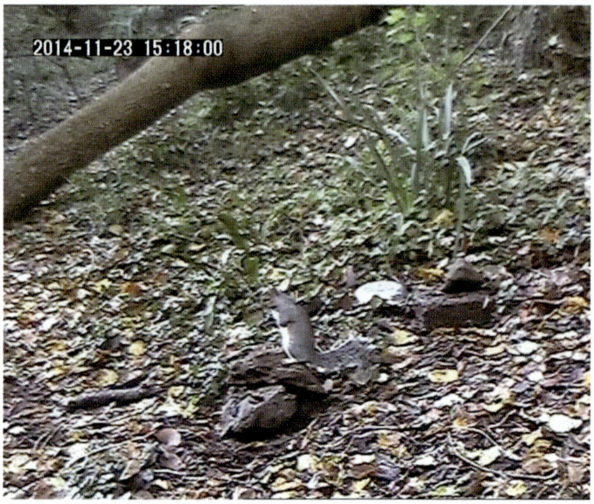

Figure 7.31 This squirrel video shows what can go wrong with quality.
https://youtu.be/ZEzmoJkDnxQ

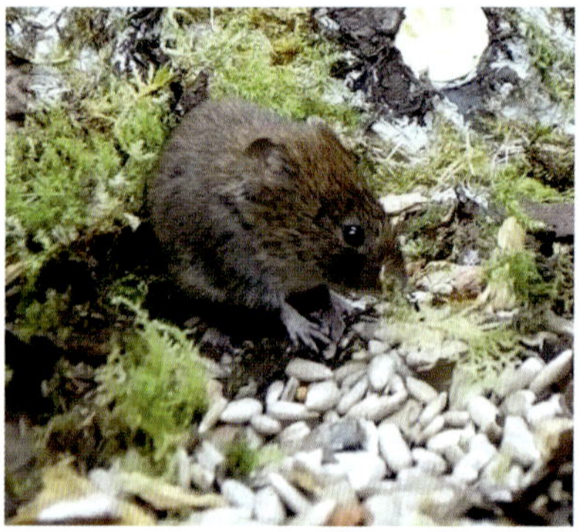

Figure 7.32 Two voles disagree over ownership of the mammal house.
https://youtu.be/IgLjBtTKdMg

8. Wildlife case studies

These wildlife case studies are intended to give a flavour of what can be done with CCTV to monitor wildlife.

- 8.1 Clean pond dipping.
- 8.2 Catching the frog catchers.
- 8.3 Rock pool surveying.
- 8.4 Badger activity.
- 8.5 Bird feeders and baths.
- 8.6 Tawny owls.

8.1 Clean pond dipping

8.1.1 Introduction

Pond dipping is a popular pastime for children, and using a submersible CCTV camera is an excellent way to show children (and adults) what life in a pond is really like. There are many advantages in pond dipping with a camera, as pond creatures can be seen behaving naturally, and need not be removed from their environment. The edges of streams and shallow rivers can also be safely explored and a range of habitats investigated. Often events missed by the naked eye can be seen clearly on a video.

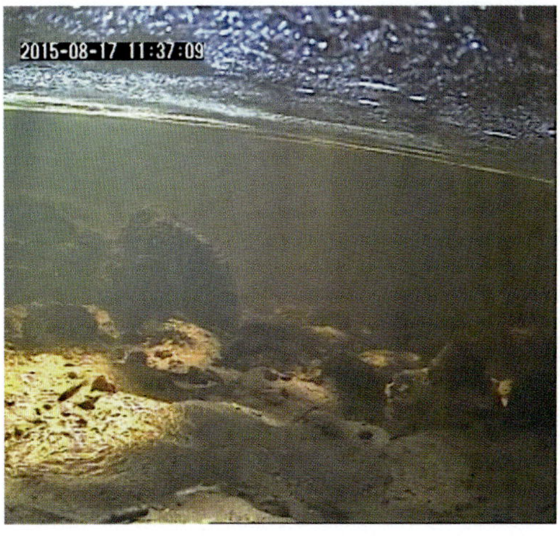

Figure 8.1 The pond is a mysterious world of light and shade.

There are several ways to carry out 'clean' pond dipping: the camera can be mounted on a rig and moved slowly through the water with the video streamed to a portable monitor, or set up as a permanent feature if pond and mains power supply are within 5–50 m of each other and the cables can be safely positioned. Although the adapter with the camera converts the mains voltage to 12 V DC, care should be taken to ensure that the mains supply is well shielded from the wet area. For example, power extension leads must not be used. For the greatest safety, the mains power connections should be inside.

Both systems were tested as described below and the advantages and disadvantages of each method are detailed.

8.1.2 Equipment

- Submersible camera.
- Recorder.
- Monitor.
- Camera support.
- Power source.

8.1.3 Method 1 – mobile system

A basic mobile pond dipping system is shown in Figure 8.2.

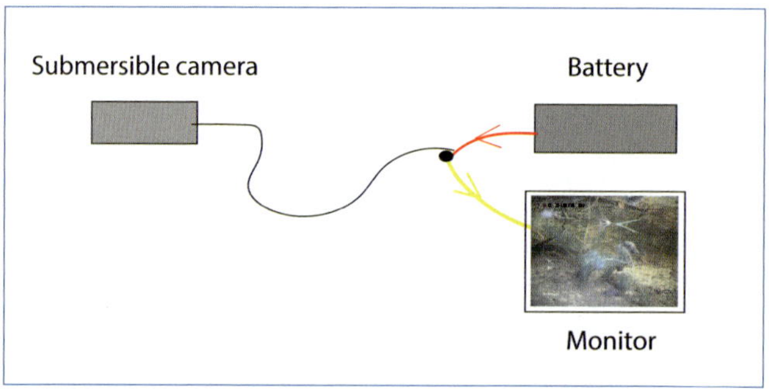

Figure 8.2 Basic outline of pond dipping method 1.

Camera

- Submersible camera.
- Lens 3.6 mm.
- Resolution 500 TVL.
- Voltage 12 V DC.
- Current 120 mA.
- Connections on the cable are female BNC for video, standard 2.1 mm barrel female connector for power.

Power source

- 8 Ah lithium polymer battery for camera.

Portable monitor

- Lilliput 5 inch monitor with integral battery.

Camera support

- Camera rig (see Figure 8.3).

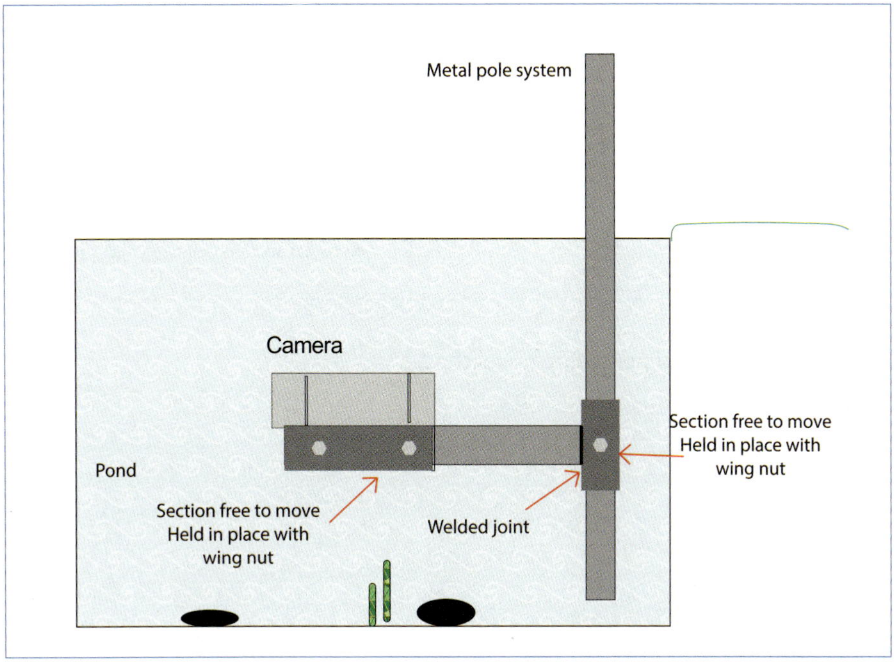

Figure 8.3 Adjustable rig for carrying the submersible camera.

The camera rig constructed for this method consisted of two square tubes, one of which was free to move up and down and was held in place with bolts and wing nuts. One end of a bolt passed through a nut welded to the metal tube and was tightened by the wing nut which was welded to the other end of the bolt. The camera was mounted on another metal tube which slid over the horizontal tube and was held in place with a wing nut assembly. The camera could be moved up and down by adjusting the wing nut assembly holding the vertical tube. This allowed for ponds of different depth. The camera was connected to the battery and to the monitor.

The camera rig was positioned in the pond with the camera about two-thirds of the way down into the pond, and the image viewed on the monitor.

8.1.3.1 Results

The system worked well in that it gave a clear image of what was in the pond, particularly at close range. The water was reasonably clear, when vision was not restricted by pond weed. However, it was difficult to keep the rig completely still. Slight movement resulted in a shaky video.

8.1.4 Method 2 – fixed system

A fixed pond dipping system is shown in Figure 8.4.

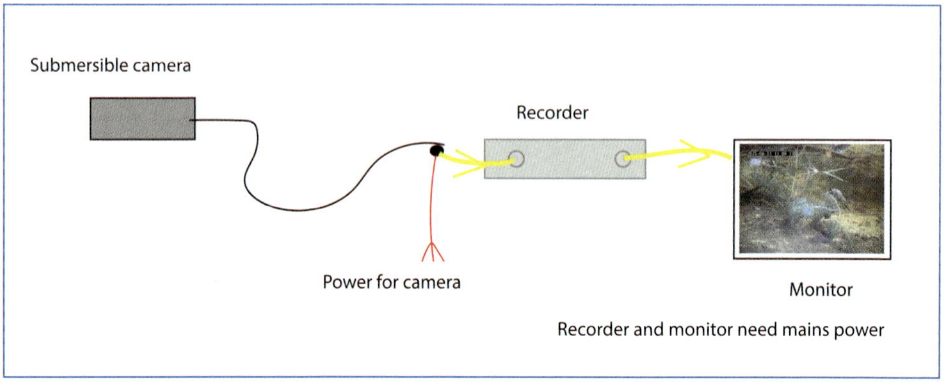

Figure 8.4 Basic outline of pond dipping method 2.

Camera

- Submersible camera.
- Lens 3.6 mm.
- Resolution 500 TVL.
- Voltage 12 V DC.
- Current 120 mA.
- Connections on the cable are female BNC for video, standard 2.1 mm barrel female connector for power.
- Mains power adapter.

Recorder (if required)

- Any analogue video recorder with BNC video inputs.
- Mains powered.
- Motion-triggered recording.
- Daytime time scheduling.

Camera support

- Ground screw.

Power source

- Mains with adapter to convert to 12 V DC for camera.
- Mains power for recorder and monitor.

Monitor

- Any monitor that accepts analogue video. Requires mains power.

For this method a fixed system was tried in two forms. In the first, the camera rig was used with a ground screw (see Figure 8.5). This was screwed into the base of the pond using a metal pole through the hole at the top of the screw.

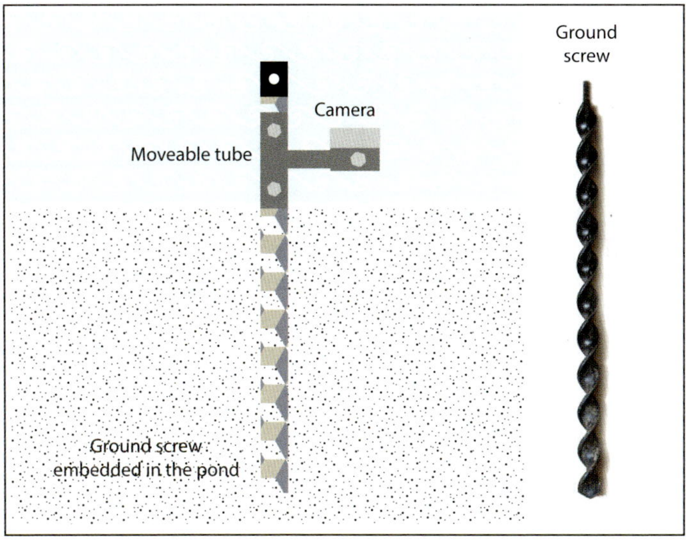

Figure 8.5 Ground screw method for supporting the submersible camera.

The second form is suitable for ponds with a liner. The camera, in its cradle, was placed directly on the floor of the pond (see Figure 8.6). This was suitable for shallower ponds where the bottom could be reached easily.

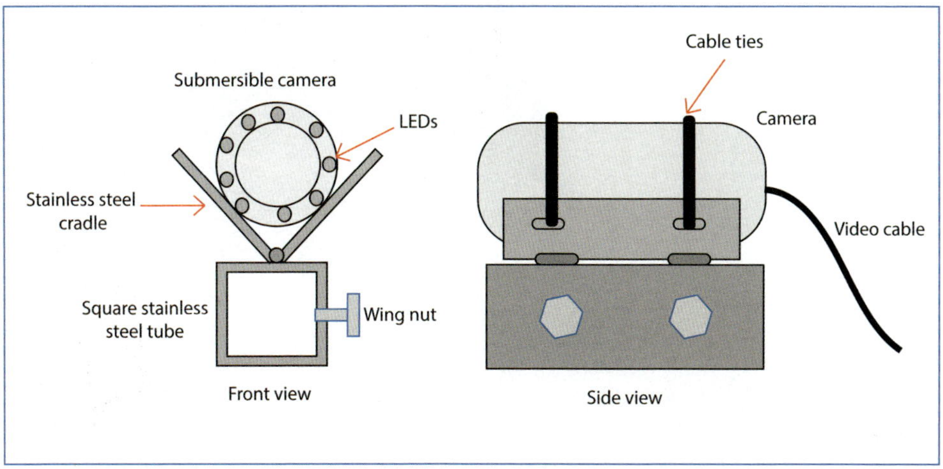

Figure 8.6 Details of submersible camera support.

8.1.4.1 Results

The fixed system gave a smooth video without the problems of hand-holding the camera rig, and is a suitable method for showing pond life when viewing indoors. If a recorder is used, then pond action can be viewed at convenient times. The disadvantage is that a cable has to be laid from the rig to the recorder and/or monitor indoors. This is not a difficult issue to address, but requires care to ensure the cable does not become a tripping hazard.

8.1.5 Conclusions

Each method has its merits, with the portable system being more suitable for natural ponds in the countryside, and the fixed system being more suitable for ponds with a mains supply close by. Educational establishments in particular would benefit from either method, especially a school pond with a fixed camera and the feed relayed to a monitor inside, with or without a recorder. If the system is to be used for long periods, a monitor would be the better option, with recording only being used at intervals for study, to prevent large numbers of videos being produced and requiring processing.

8.2 Catching the frog catchers

8.2.1 Introduction

February is the main spawning period for frogs (*Rana sp.*) for much of the UK, during which they are more vulnerable to predators, and the close contact can spread disease.

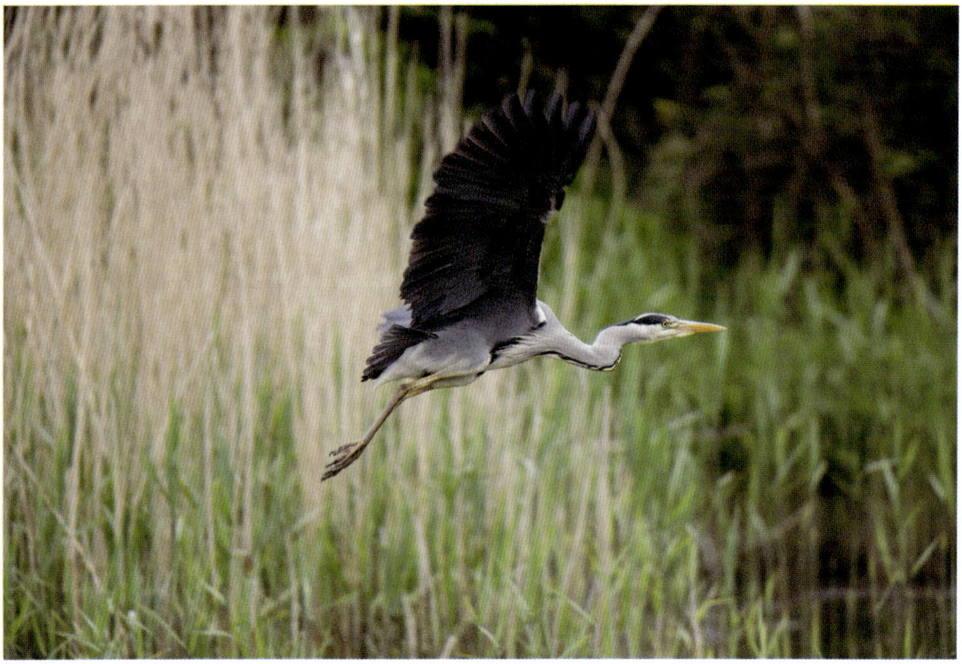

Figure 8.7 Heron 'frog catcher'.

The last few years have produced headlines to the effect that *Ranavirus* is killing off the UK frog population, but a more recent report by the British Trust for Ornithology (Humphreys *et al.* 2010) suggested that populations have stabilised. Even with the threat of virus apparently reduced, frogs are susceptible to predators, with the grey heron (*Ardea cinerea*) being the main suspect, although other birds and mammals are also suspects. Herons are very sensitive to disturbance, and a thin film of oil on the pond surface is often the only sign that a large bird has been in the water.

Although articles and books say that frogs have many predators, there is little detail on the likelihood of a frog being taken by any particular predator. CCTV seemed to me to be a good method of obtaining a little data on the subject, and I used one of my wildlife ponds as the testing ground. The pond is in a rural part of south Devon and is surrounded by wildflower meadows and trees.

8.2.2 Equipment

- CCTV camera.
- Lead acid battery.
- Tripod.
- SD-DVR recorder.
- Test monitor.
- Box.

The basic equipment setup is shown in Figure 8.8.

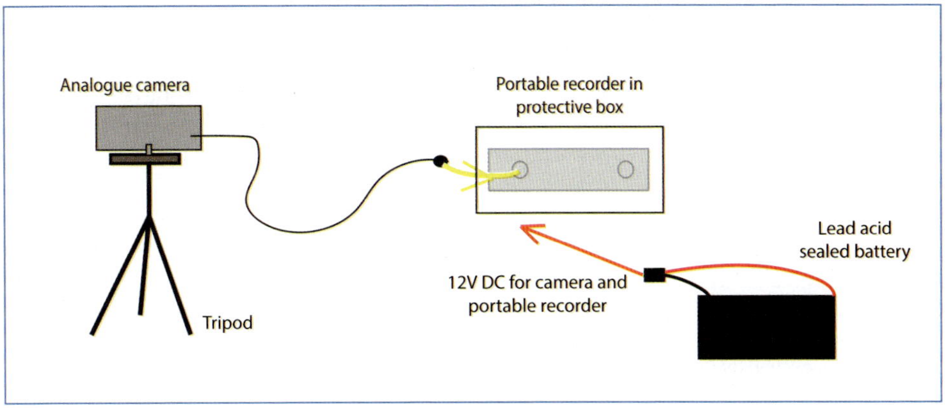

Figure 8.8 Basic equipment setup for frog catchers.

Camera

- Analogue day/night infrared camera.
- 630 TVL.
- Varifocal lens 2.8–12 mm.
- Voltage 12 V DC.
- Current 550 mA.
- Infrared LED range 30 m.

Battery

- 110 Ah sealed lead-acid battery. A lithium battery could also be used.

Recorder

- Genie SD-DVR.
- Motion recording sensitivity at medium.
- Scheduled for 24 hour motion recording.

Box

- Small Peli case to contain the recorder.

Camera support

- Tripod.

8.2.3 Method

As no power was available next to the pond, I used a 110 Ah lead-acid sealed battery. A sealed battery is safer for use as it prevents acid spills. It was described as a 12V leisure battery suitable for electronic equipment. Although lead-acid batteries can be used to supply power to a CCTV camera and recorder, care must be taken not to let the battery discharge by more than 60%. Even batteries suitable for a deep discharge do not give a constant 12 V as the battery discharges, and the voltage can drop quite quickly.

The voltage of the battery was checked using a voltmeter each day and I found that after 6 days, the voltage had dropped to 11.8 V and started to fall very quickly thereafter. In this case 6 days was the maximum time the battery could be used before charging and was a convenient period for a pilot study.

Making electronic connections to a lead-acid battery is difficult simply because the connections are so different. Connectors supplied with batteries have very thick wire, and this is one situation where ready-made leads are hard to find and leads are best made up, as shown in Figure 8.9, with the connection details described earlier in section 7.5.

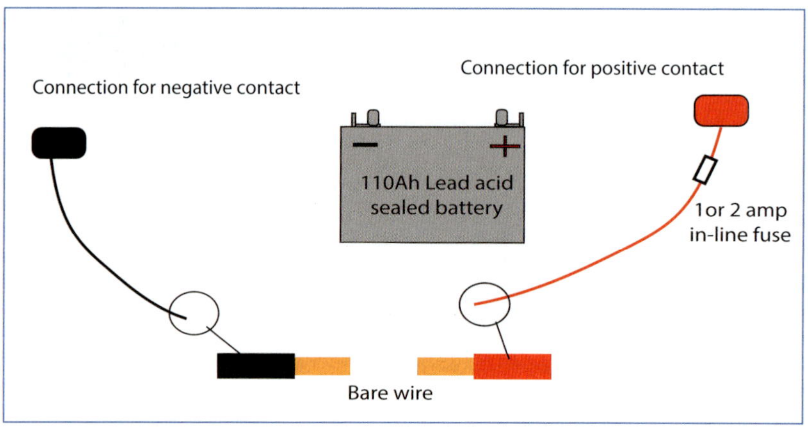

Figure 8.9 Battery terminal connectors.

Battery connectors

The best connectors are leisure battery push on connectors. These have plastic insulating covers and are colour coded for safety.

Cable

Suitable cable is 6 mm^2 in cross-section. Separate red and black wire should be purchased. This (or similar) can be found in an auto shop or on eBay. The cables should be connected by the screws of the appropriate battery connector.

Fuse

An inline or blade fuse should be connected to the red positive wire and should not be rated at more than 2 A.

Terminal block

The wires from the battery could be screwed to a terminal block, taking care to match positive to positive and negative to negative. The terminal block should be enclosed for safety. Details of the connections are shown in Figure 8.10.

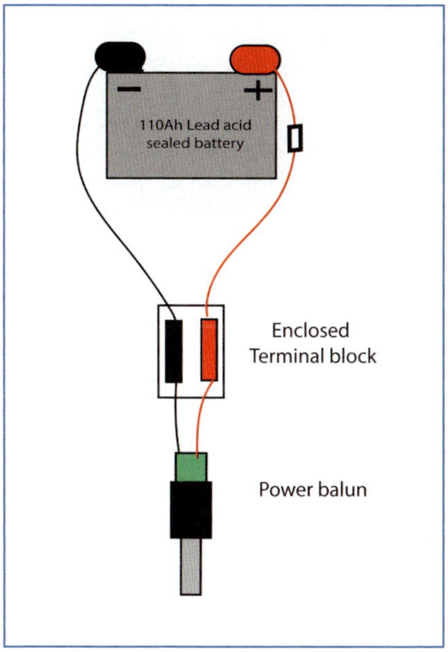

Figure 8.10 Connecting the battery to a suitable sized connector.

Power balun

The other side of the terminal block should be connected to thinner wire, which in turn should be connected to a power balun (see Figure 2.9 for a description of a power balun).

110 Ah batteries require treating with care and wiring should not be carried out without previous experience. If the positive and negative wires touch this will short out the battery and can lead to sparking or overheating. The fuse is a precaution in case the connections accidentally short.

Carrying the battery

As the shed hide was in a 5 acre garden, a wheeled sack trolley was used (see Figure 8.11). Simply placing the battery on the carry plate of the trolley did not work, as I soon discovered the slightest bump made it bounce off. Even smooth surfaces were a trial, as slight slopes caused the weight of the battery to pull it to one side. The solution was to put the battery in a large plastic box (Really Useful Box 35L). The box was fixed to the trolley with tight bungee cords. This has the added advantage of keeping the battery connections safely inside the box and not trailing around.

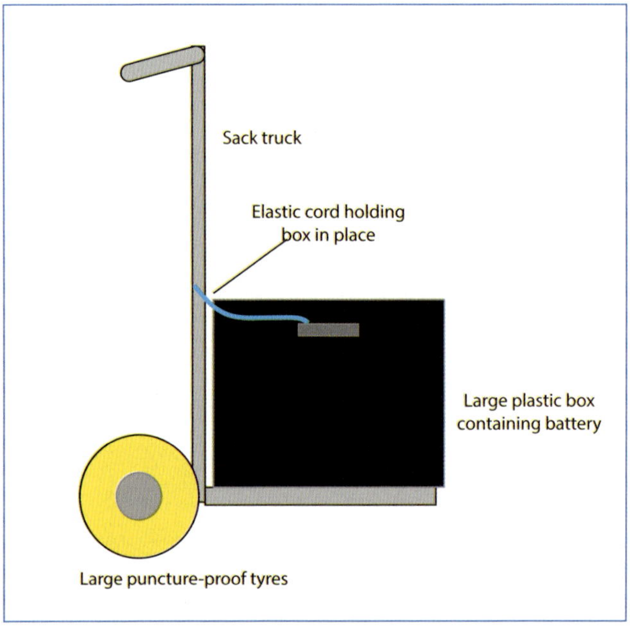

Figure 8.11 Transporting the battery across rough ground.

Fixing the camera

The camera was fixed to a tripod using my fixing rig comprising a Manfrotto tripod adapter and a piece of exterior plywood (a detailed diagram of the adapter can be seen in Figure 8.23). The camera was attached to the battery by way of the power balun which also powered the recorder using a DC power splitter cable. A test monitor was required to ensure that the system was working properly and that the camera was pointing in the right direction and was focused.

8.2.4 Results

After several days and nights of motion-triggered recording, it became clear that the main frog catcher was a heron. Time after time it was filmed catching and eating those frogs too engrossed in their mating to escape. Most of the hunting took place at dawn and dusk.

However, the heron was not the only bird to favour the occasional frog. A tawny owl (*Strix aluco*) also appeared several times. Initially it seemed disturbed by the infrared LEDs on the camera which emit a very dull red glow, and flew off almost as soon as it arrived. Eventually, after a few trial runs and a spell perched on a tree, it ignored the LEDs and appeared to catch a few frogs of its own (see Figure 8.12).

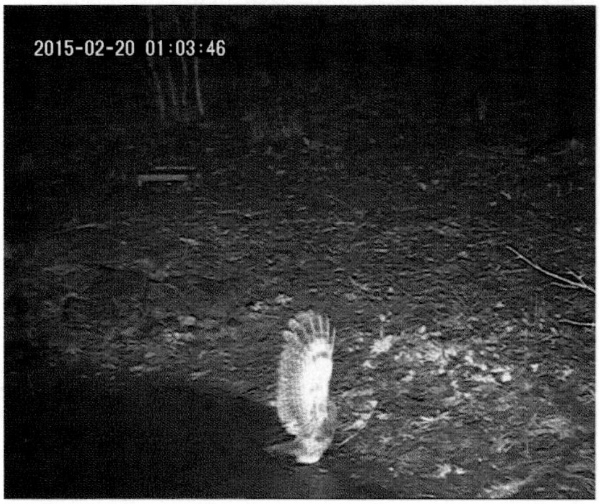

Figure 8.12 Owl frog catcher.

Predation of the predator

On one occasion the heron itself looked in danger from a predator when a fox appeared on the scene. Although busy frog hunting, the heron was alert enough to see the fox arrive and flew off. The fox sniffed at the pond, but did not appear interested in the frogs and wandered off after a scratch.

Longer monitoring periods would give more information on frog predators, and CCTV is a suitable method, especially if mains power is used. Trail cameras could be used, but the short videos and lack of a pre-record feature would mean discontinuity in the gathering of data, and the chance that short-lived events would be missed.

8.3 Rock pool surveying

8.3.1 Introduction

There is a wealth of information to be found in rock pools (see Figure 8.13) and the standard rock pool surveying technique for the amateur naturalist is to walk through or next to pools, lift up stones to see what is underneath, or catch creatures in a net. This method is used in particular by children and it is possible that this method of rock pooling is predominantly seen as a child's occupation.

The professional biologist has several survey techniques: netting, visual assessment or the addition of an anaesthetic to the pool. This last method is generally used for counting fish and the effect on other creatures is unknown (Davies *et al.* 2001).

Each of these techniques has disadvantages: moving stones disturbs the environment in the pool; netting disturbs the pool inhabitants; visual observation is inaccurate as

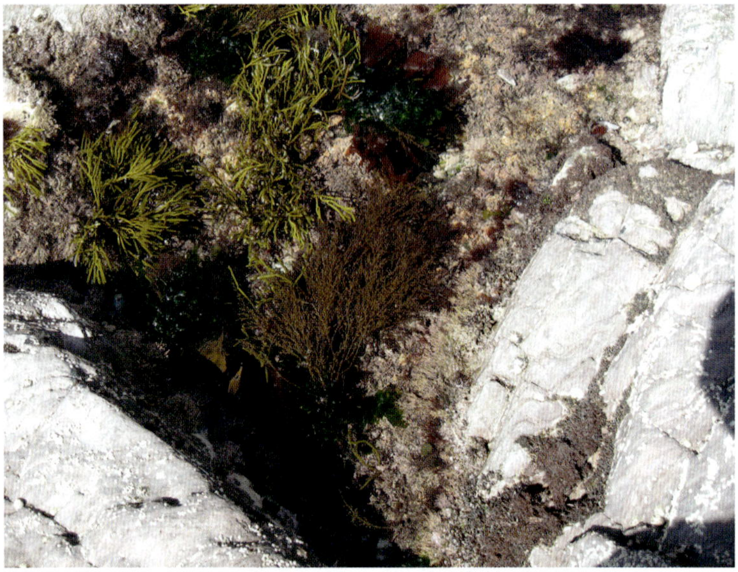

Figure 8.13 The rock pools at Wembury contain crystal clear water.

pool creatures tend to hide, and are very sensitive to the presence of an observer; use of anaesthetics is unpleasant and has an unknown effect on other pool inhabitants.

I considered that a submersible CCTV camera would be an addition to these techniques with fewer disadvantages and tested the system in the rock pools at Wembury in Devon.

8.3.2 Equipment

- Submersible camera.
- Camera rig.
- Battery.
- Recorder.
- Ground screw.
- Test monitor.

The basic setup for rock pool surveying is shown in Figure 8.14.

Camera

- Lens 3.6 mm.
- Resolution 500 TVL.
- Voltage 12 V DC.
- Current 120 mA.
- Cable length 5 m or more.
- Camera rig as in Case study 8.1

Battery

- 8 Ah lithium polymer battery.

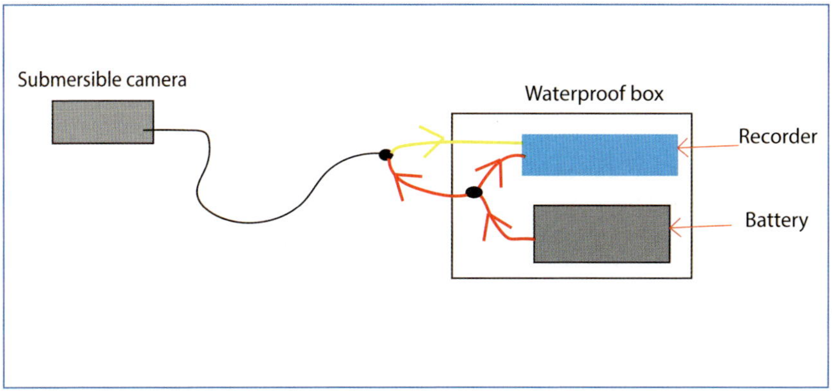

Figure 8.14 Basic equipment setup for rock pool surveying.

Recorder

- Genie SD-DVR.
- 32 GB SD card.
- Recording set to motion detection.

8.3.3 Method

The most important information required for rock-pooling is knowledge of the tide times as it is easy to get caught by the tide. As I wanted as much time as possible, I decided to start with rock pools close to shore and move outwards as the tide receded. Low tide was at 14.20, so I began at 11.20 and started to return back to shore at low tide. This gave a window of 3 hours, although this could be extended to 6 hours as long as care was taken to move back to shore gradually and to keep a close eye on the tidal progression.

The rock pool area was large, and it was easy to stay within sight of the beach for personal safety and still be far enough away from other rock-poolers who might disturb the wildlife being filmed.

The first thing I noticed was that each time I was about 2 m from a pool, fish darted under seaweed and did not return as long as I stood and watched, highlighting just how sensitive wildlife is, even at sea.

There were a large number of pools of varying sizes and depths, but each one had an extensive range of colours and textures of sea weeds. Limpets and sea snails abounded, but there seemed to be very few living creatures, no doubt hiding after my approach.

In the first pool selected I managed to fix a ground screw into the base of the pool, which was more shingle than rock. I attached the camera and connected the recorder and battery. I used my CCTV test monitor to ensure that the setup was working and giving a clear image.

The camera gave a very clear image with intense colours that were very true to reality. The absence of water movement, other than a slight windblown surface ripple, meant it was possible to see right across the pool.

After 30 minutes, I moved the camera to another pool further away from land, but this time, as the pool was shallow, I placed the camera in its holder directly on the bottom of the pool. I left this for another 30 minutes and repeated the process.

Throughout this process I saw no moving life other than the fish darting into hiding as I approached. The rocks were covered in sea snails and limpets.

8.3.4 Results

The videos showed a great deal more than my visual assessment. Once I had moved out of sight, and peace had been restored to the rock pool, tiny fish could be seen swimming in and out of the seaweed. Rock pool fish predated on tiny shrimps. Most unexpected was how far apparently stationary limpets and sea snails moved around as they grazed on the rocks. Comparing the two images in Figure 8.15 and Figure 8.16 shows how far they have moved in 1.5 minutes. This is shown dramatically if the video is speeded up.

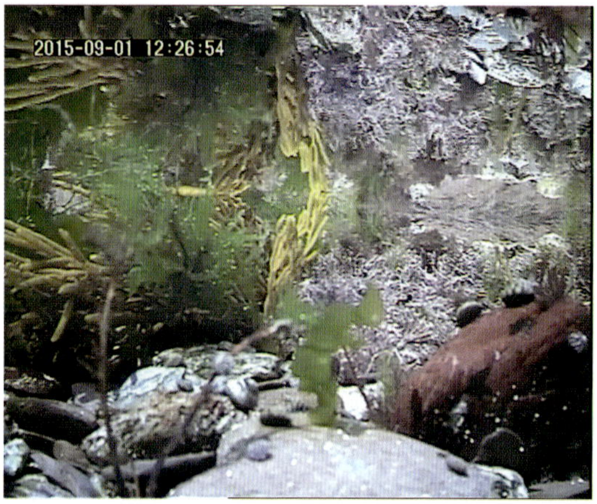

Figure 8.15 Sea snail and limpet initial position.

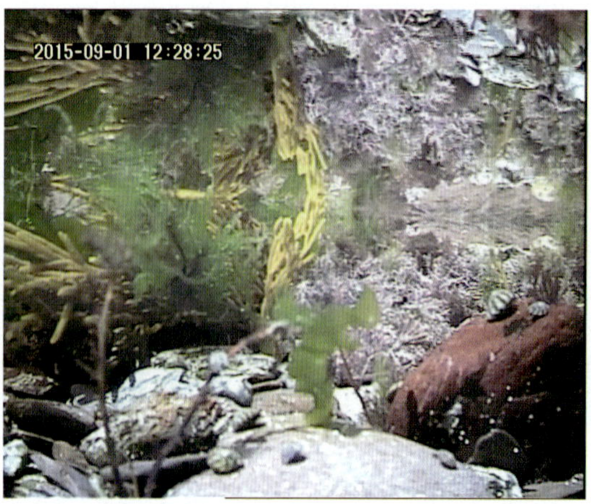

Figure 8.16 Sea snail and limpet position after 1.5 minutes.

8.3.5 Conclusions

The submersible camera proved to be an excellent tool for monitoring rock pool life. Clear images in true colour showed facets of rock pool life that more invasive techniques would not capture. In addition, the procedure caused little disturbance to rock pool life and allowed natural behaviour to be committed to video.

The use of a submersible camera is a compromise technique that I believe is of value to both the amateur and the professional marine surveyor. It does not disturb the rock pool environment or its inhabitants, can 'observe' for long periods and provides a record to show the behaviour of pool inhabitants.

Short videos of rock pool action would also have educational value and videos could be shown in coastal marine centres or schools.

The equipment list can be simplified as shown:

- The battery will only be used for a few hours at a time for each survey and 4–8 Ah is more than adequate. For short visits (cheaper) batteries with lower capacity could be used.
- Short cables of 3–5 m are all that is required in most cases.
- Although a ground screw can be useful in deeper pools with a shingle base, the camera can be used with just its support stand in pools with a rock base.
- The motion of seaweed and rising air bubbles can give rise to constant recording, so a lower specification of recorder without motion detection could be used.

8.4 Badger activity – a life story

8.4.1 Introduction

The thrill of seeing a badger in the wild is hard to beat. However, watching badger behaviour throughout the year is not only exciting, it can also yield a great deal of

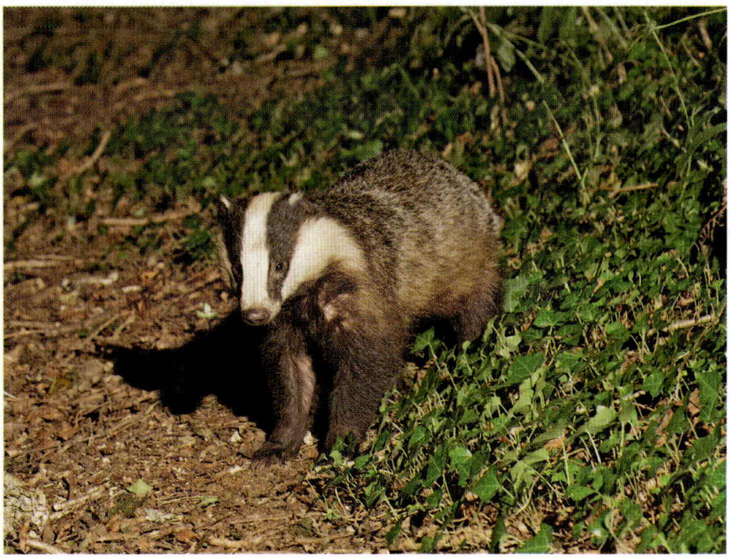

Figure 8.17 A floodlit badger (*Meles meles*) in woodland.

information about badger interaction. Trail cameras are good for short events, but as most trail camera videos only last 60 s at most, they are arguably more useful as an event monitor than a method for studying behaviour. Many 'badger watch' viewing events can be found, often in holiday areas, with a shed with power acting as the viewing hub, and this can be emulated using CCTV.

I described earlier how power can be brought to a shed, but if you live in a rural area, badgers are quite likely to come to you. Patches of scratched up grass or holes dug for toilets are signs that badgers are nearby. This case study describes how to set up a viewing hub and what might be seen. In this example I am assuming that mains power is available.

8.4.2 Equipment

- Camera.
- Recorder.
- Monitor.
- Tripod.

A basic setup for badger monitoring is shown in Figure 8.18 and Figure 8.19.

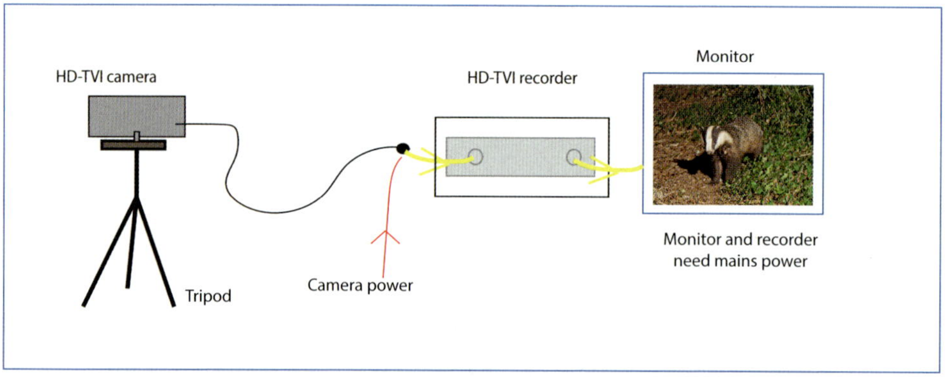

Figure 8.18 Badger monitoring basic set up.

Camera

- HD-TVI day/night camera.
- Resolution 1,920 × 1,080 pixels at 25 frames per second (fps) or 1,280 × 720 pixels at 50 fps.
- Lens 2.8–12 mm.
- Voltage 12 V DC.
- Current 650 mA.
- Waterproof to IP68.
- 12 V DC mains adapter.

Power

- Mains.
- Floodlight (optional).

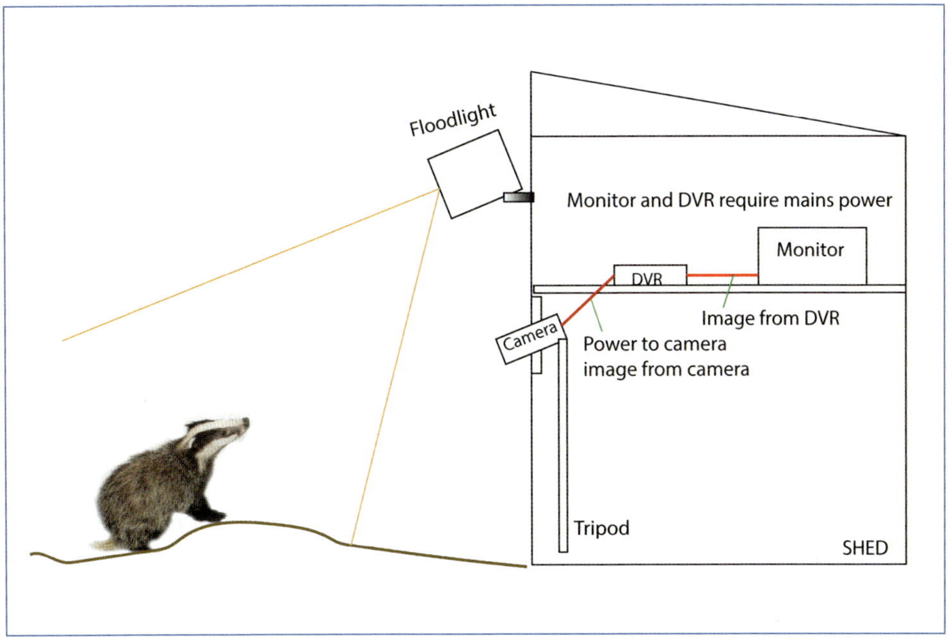

Figure 8.19 Badger monitoring.

Recorder

- HD-TVI recorder with 1 Tb hard disk and four inputs (only one used).
- Recording set to highest resolution.
- Scheduled motion detection from dusk to dawn.
- Pre-record at 10 s.
- Mains power.

Camera support

- Tripod with camera rig (see Figure 8.23 for details).

Monitor

- Full size monitor.
- Mains power.

8.4.3 Method

The camera was fixed to the tripod using the camera rig. As power, camera and recorder were close to each other, the camera was attached directly to the recorder without additional cables. The recorder was attached to a monitor, in this case an old PC monitor. The monitor only supplied a VGA input for the video signal, but the recorder had a VGA output as well as an HDMI output. The fact that the monitor had a VGA (non-HD) input did not affect the recording quality, which was in HD.

The recorder used was an HD-TVI recorder and was very easy to set up. This is not always the case with recorders. It was easy to select each day and schedule the times at which motion detection was to take place. The sensitivity of motion triggering was variable, but low sensitivity was used as badgers are large animals and do not require more sensitive motion detection. Because a CCTV camera was being used, filming was continuous, and recording took place only when motion within the image triggered the recorder to record. As the recorder was not a mobile recorder, it was fitted with a hard disc. This greatly increased recording capacity.

Night vision HD-TVI CCTV cameras give very high-quality images at night, without the grainy effect that can be seen in cameras of a lower specification.

In my experience, badgers are not disturbed by floodlights. I have a video of a badger eating peanuts when the floodlight came on. It stopped for a second, looked around, and then carried on eating. LED floodlights are now available, together with suitable brackets, and are worth investing in if you wish to film badgers in colour at night; they use less electricity than halogen lighting.

8.4.4 Stages to look out for

I have been monitoring badgers near me for years, and there are several interesting stages to look out for. Although badgers live in social groups they tend to forage on their own. I have a female badger that has been coming for peanuts near the badger hide for some time. At one point in the year a male badger comes along too. Although badgers can mate at any time, the peak times are in spring. The male seems to stamp his authority by shoving the female out of the way when he spies the peanuts. Pre-mating and mating behaviour is rough. A video still is shown in Figure 8.20.

Once the mating period is over, the male disappears. In early summer any cubs (usually one in the case of my local badger) are shown round the peanut area. At other times badgers visit, some from the same social group, some not, and some fascinating encounters can take place.

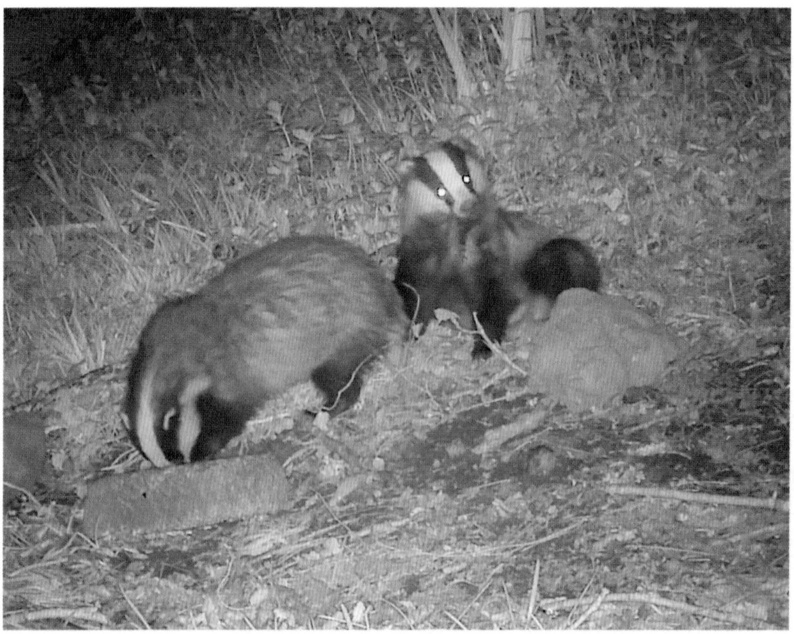

Figure 8.20 A female badger surveys the scene while last year's cub forages.

8.5 Bird feeders and bird baths

A log feeder complete with visitor is shown in Figure 8.21.

8.5.1 Introduction

Most of us like watching the action at the bird feeder, but there is always the chance we will miss a bird or event of significance. Using CCTV to monitor the feeder will ensure that nothing is missed. The advantage of feeder cams is that they can be set up near to the house and are likely to have access to mains power and the internet.

There have been many occasions when I wished I had set up a camera, for example when a blue tit and long-tailed tit had a disagreement, or when a great spotted woodpecker brought its youngster along to show him or her the ropes. In winter I have had redwings, pied wagtails and blackbirds competing for apples, plus a myriad of other encounters, with weasels and foxes occasionally getting in on the action.

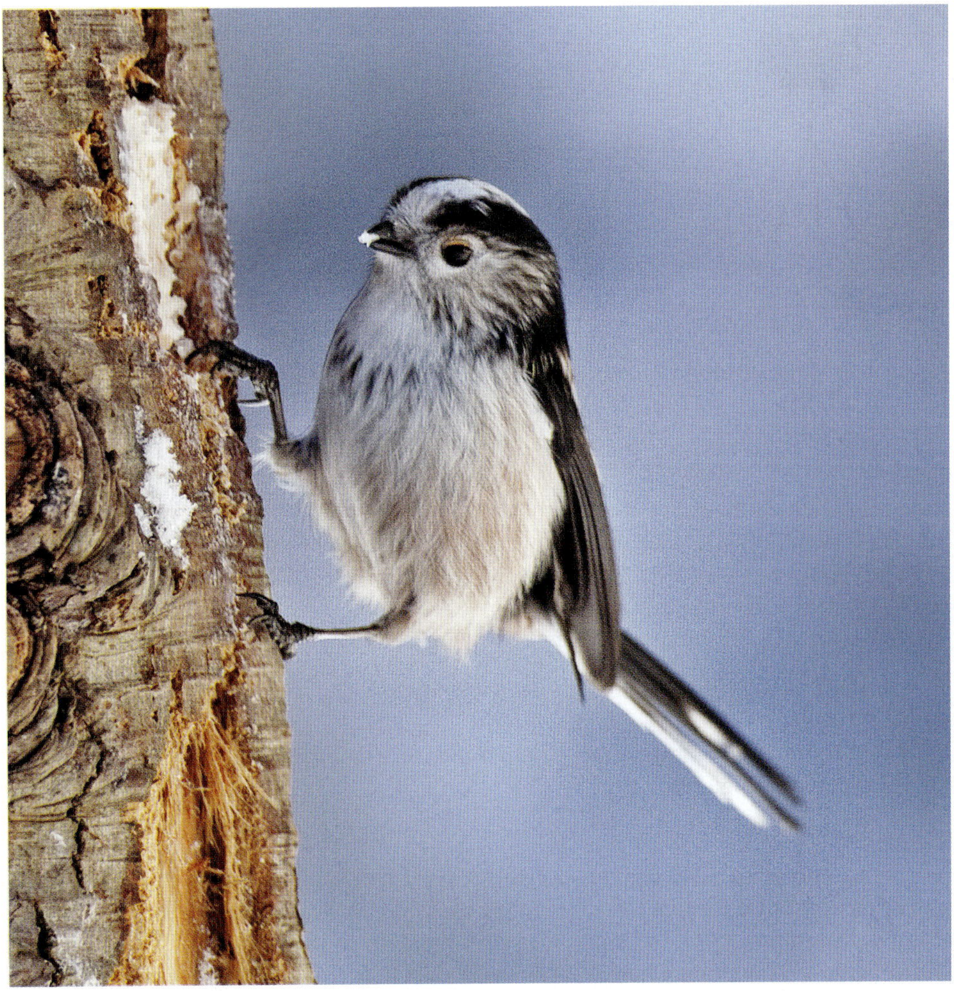

Figure 8.21 A long-tailed tit (*Aegithalos caudatus*) on a log feeder.

Many people set up feeder cams which show the action on traditional nut and seed feeders, but a technique used in wildlife photography is to set up log feeders, which make more attractive subjects for feeder cam videos. In addition, a camera can be set up near a bird bath, which if it is natural looking, can also supply excellent subjects.

8.5.2 Equipment

- Camera.
- Recorder.
- Monitor.
- Camera support.

A basic setup for filming bird bath action is shown in Figure 8.22.

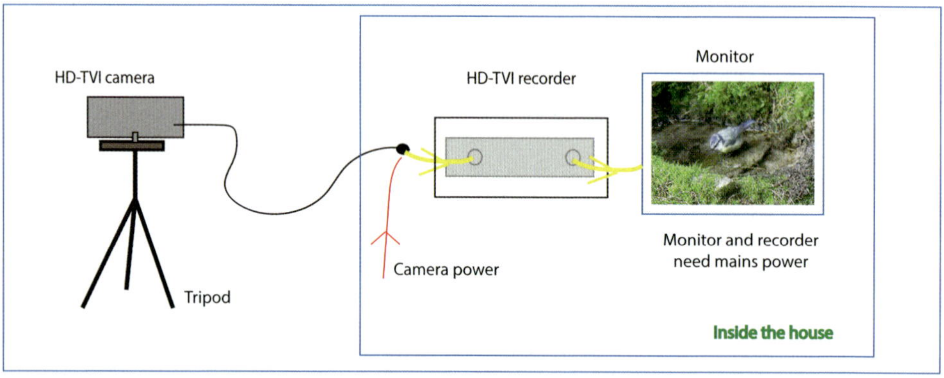

Figure 8.22 Bird bath and log feeder basic setup.

Camera

Assuming mains power and the internet are available, there is a wide range of cameras to choose from. HD-TVI and IP cameras could be used. For an HD-TVI camera, use an HD-TVI recorder and for an IP camera, use iCatcher software or the camera manufacturer's own software for recording.

I used an HD-TVI camera with the following parameters:

- Lens with a variable focal length of 2.8–12 mm.
- RG59 cable.
- 12 V DC from a mains adapter, plugged in inside the house.
- Immersion rating IP68.

Recorder

As bird feeders see a great deal of action, it may be advisable either simply to watch the action via a monitor, or live stream from the internet to watch it remotely. Recording is also possible, although it will mean many long videos to trawl through. However, near a bird bath, a motion-triggered recorder is suitable, as there is much less action and a conveniently manageable number of videos produced.

Monitor

Watching bird feeders on a monitor is the easy way to view feeder action, when either you cannot view directly, or a number of people wish to view. For example, a feeder monitor is very suitable for schools, where it could become part of a lesson. There is a wide range of monitors, but I have found that almost any PC monitor will do. These can be picked up cheaply second hand, but an adapter/convertor may be required to convert the video signal to one accepted by the monitor.

Camera support

CCTV cameras are often designed to be fixed to a wall or shed, but for a flexible portable support, I have designed an effective device using a small piece of exterior plywood and a tripod adapter. The tripod adapter is screwed into a drilled hole in the plywood, and a large hole is drilled to pass the camera cable through. The camera is screwed onto the plywood as shown in Figure 8.23, and on the opposite side to the adapter which is then fixed onto the tripod.

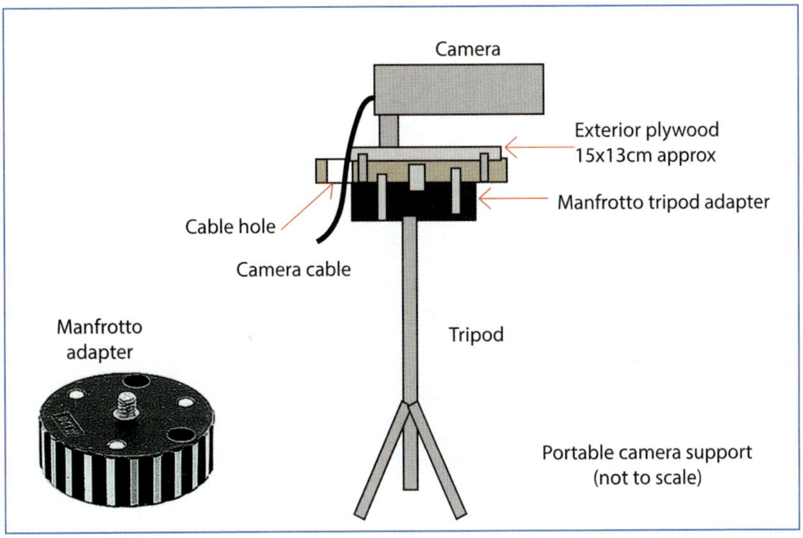

Figure 8.23 Details of the tripod camera support.

This very simple method works very well and allows easy adjustment for the camera angle, especially if the tripod has a ball head. If the ground is soft, the plywood support can be bolted to a ground screw. This is very suitable for filming action near a bird bath.

8.5.3 Method

8.5.3.1 Log feeder

A log feeder requires a log about 5–10 cm in diameter with holes sloping at an angle of around 45° below the horizontal, drilled in along its length from top to bottom. This stops the filling from falling out. The holes are filled with a mixture of lard and chopped nuts. Woodpeckers are particularly fond of this mix, but smaller birds will also visit the log feeder. Log feeders make much more attractive and natural subjects than standard feeders.

The tripod support should be placed near the log feeder. Ideally it should be camouflaged, as I have found that birds quickly become used to the sight of tripods, and frequently use them as perches. Camouflage can prevent your equipment being obscured by bird lime. The cable should be brought into the house, either through a hole in a wall or window frame, or temporarily through an open window.

8.5.3.2 *Being ready for the unexpected*

Some 'feeders' are natural and it is a good idea to have the equipment at your fingertips, ready to go, when the unexpected happens. A crab apple tree in my garden became the focus of attention of a small flock of redwings (*Turdus iliacus*) one very snowy winter. The

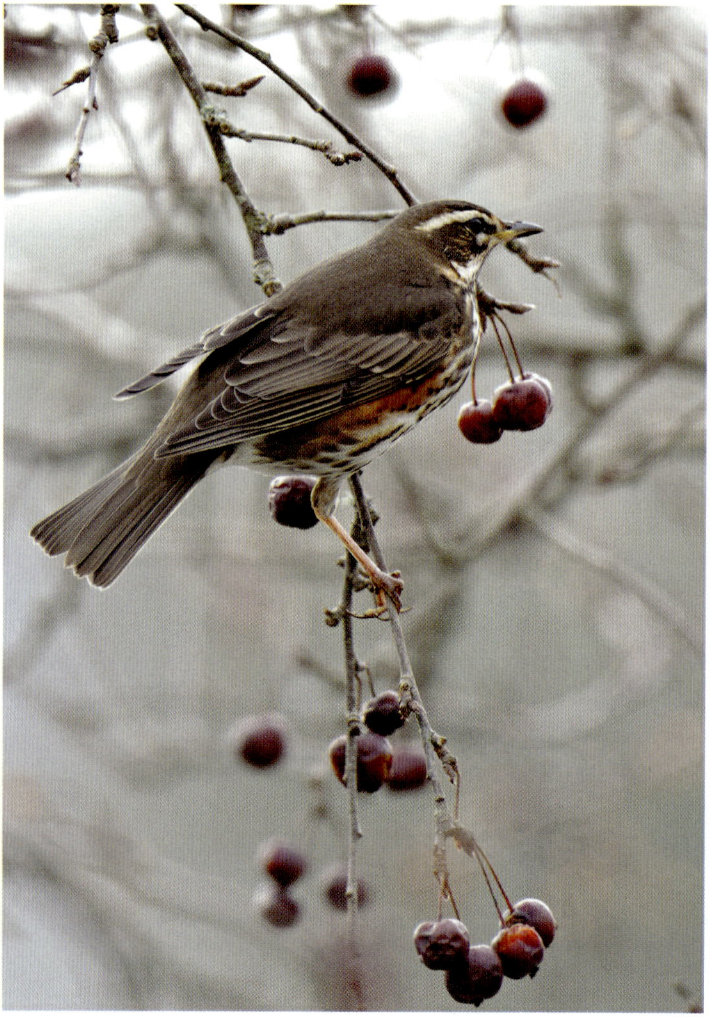

Figure 8.24 Redwing on a crab apple tree.

birds stayed on the tree from dawn to dusk for 3 days until every crab apple was eaten. I managed to get some still images (see Figure 8.24), but a CCTV video would have given a better idea of the action, which was very entertaining.

8.5.4 Bird bath action

Birds can spend most of the day on a feeder, but visits to the bird bath for a drink or a bath are fewer and easily monitored using motion detection (see Figure 8.25). Changes in light can cause triggering, so a bath in shade will give fewer 'empty' videos. As well as birds, mice, shrews and other small creatures will come to drink.

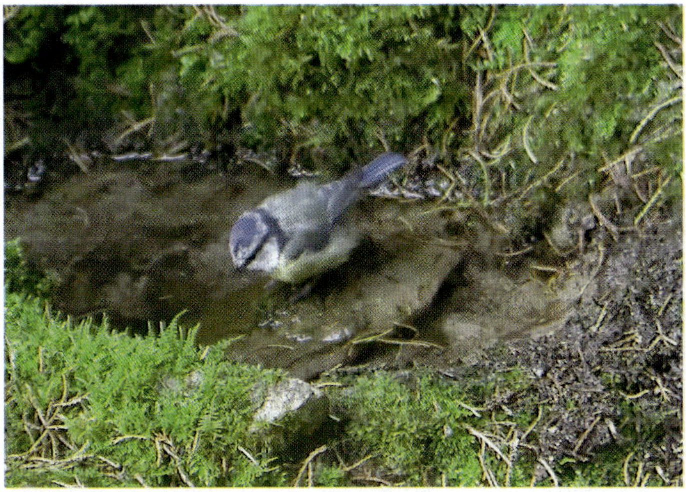

Figure 8.25 A blue tit (*Cyanistes caeruleus*) contemplates having a bath.

8.5.4.1 Internet streaming

If you have an IP camera connected to the internet, you can stream videos of your feeder action via the internet. Please read the sections on internet streaming in Chapter 4 for more detail. Because of the huge amount of data in videos, it is costly to stream large videos, so they are usually reduced in size. The monthly cost to a streaming service will depend on the pixel dimensions you want your video to have. A video of pixel size 426 × 240 will take up less bandwidth and be cheaper to stream than a video of pixel size 1,920 × 1,080, for example.

8.5.5 Conclusions

Bird baths and bird feeders, especially if natural looking, are excellent subjects for CCTV monitoring, especially in HD format. As a large number of videos can result from feeders, live viewing on a monitor or live streaming on a PC may be a more practical method, with videos being recorded on occasion. Cameras near a bird bath record fewer events and are more suitable for recording.

8.6 Tawny owl

8.6.1 Introduction

Tawny owls in my area of the UK are known to hunt on the woodland edge, and large branches close to rides are a favoured vantage point, especially if log piles are nearby. The relatively bare ground on the edge of woodland makes vole spotting easier, and as log piles are often used by small mammals for cover, their presence increases the likelihood of a successful hunt.

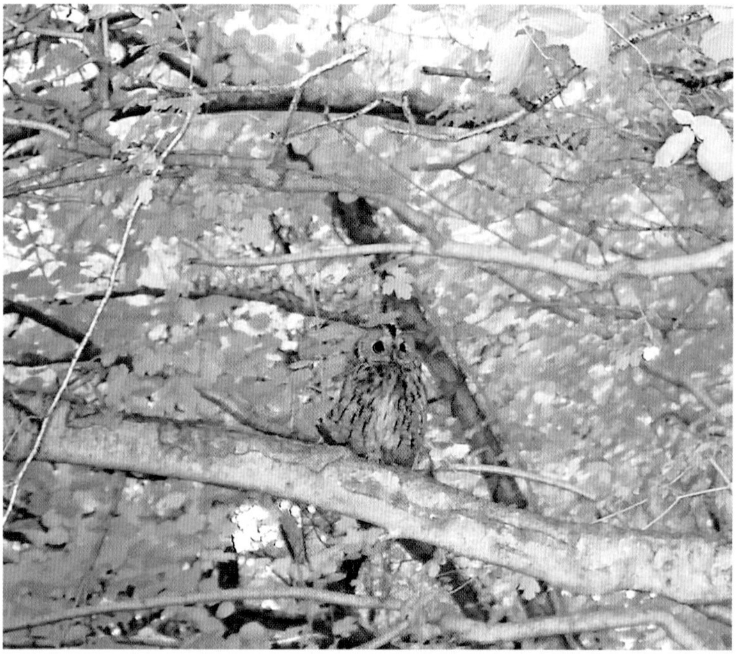

Figure 8.26 Tawny owl (*Strix aluco*) viewed from the hide in infrared using an HD-TVI camera.

The large branches on the edge of a wood I know overhang a patch of bare ground dotted with log piles, and this seemed a good position to set up a CCTV camera. The wood was known to have a resident pair of tawny owls, with a buzzard pair in the neighbouring wood. I had often seen the creamy white bird faeces associated with birds of prey under the branches which confirmed that birds of prey used them. The presence of a hut with mains power made this spot suitable for an HD-TVI camera.

8.6.2 Equipment

- HD-TVI camera.
- Compatible DVR.
- Monitor.
- Tripod for the camera.

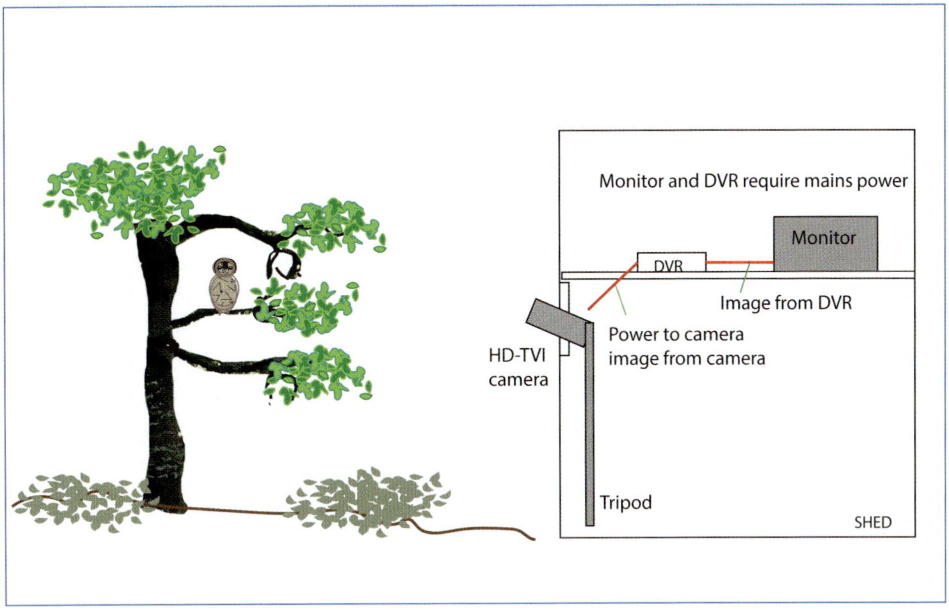

Figure 8.27 Tawny owl monitoring.

The setup for viewing tawny owls is similar to that for the badger hide and is shown in Figure 8.27.

Camera

- HD-TVI camera.
- Variable-focus lens adjusted to 12 mm (maximum).
- 12 V DC power supply.
- Connected directly to recorder.
- Tripod with camera support.

Recorder

- HD-TVI recorder.
- Pre-record.
- Motion recording.
- 24 hour scheduling.
- USB drive for exported video files.
- Masking.

Monitor

A monitor is required for setting up the angle and position of the camera.

8.6.3 Method

Overhanging branches close to the shed hide were selected as potential perches that the owl might use. They overhung some bare earth and were close to log piles. Power was supplied to the camera, DVR and monitor as shown in Figure 8.27. The camera angle was adjusted so that it pointed to the overhanging branches. Using the monitor for viewing, the camera zoom and focus were adjusted so that the closest branch was well positioned in the frame, and was in focus.

Motion detection was selected on the DVR with 24 hour scheduling. Although owls are nocturnal, the local tawny owls were first heard at sunset, and sometimes during the day. In addition, buzzards were known to use the same branches as a perch, and 24 hour scheduling meant they might also be recorded.

There were several areas of foliage that moved even in light wind, so these were removed from motion detection by using the masking feature on the recorder.

8.6.4 Results

The masking and motion recording feature worked well, and only a few videos were recorded in the first three nights, these being of moths and bats. On the fourth night at dusk, owls were recorded flying up to and sitting on the main branch, as well as similar branches further from the hide.

The owls were present about one night in four. On the second occasion, a pair of owls alighted on the branch nearest the camera. After some initial consideration, they started preening each other.

The owls invariably appeared at dusk and unfortunately just when light levels were very low but often not quite low enough to switch on the infrared LEDs. This resulted in videos that were more grainy than usual. The next step would be to try floodlights.

8.6.5 Conclusions

This study illustrated the importance of understanding the preferences of a chosen subject, as the information that tawny owls often perched on branches overhanging a woodland edge proved correct.

Due to the short period of motion, the recording of videos is a more suitable method for monitoring than live viewing or live streaming for 24 hour periods. If a pattern is established with the owls appearing at regular hours, live viewing and/or streaming would be fruitful.

8.7 Videos

Pond dipping

A video showing fish in a 'pond' is shown in Figure 8.28. This pond is actually the shallow, relatively still water at the edge of a river. Small fish can be seen swimming about, some appearing almost transparent. Pond dipping videos can be interesting if the surface of the water is seen, as it often shows reflections.

Frog catchers

The heron in Figure 8.29 spent a week catching frogs, mainly at dusk and dawn.

The video in Figure 8.30 shows the heron becoming agitated. It soon becomes apparent that it has seen a fox. The heron flies off and the fox appears casually on the scene.

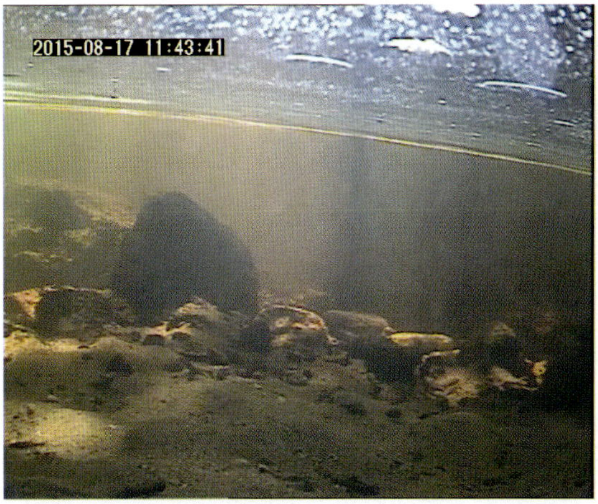

Figure 8.28 Small fish and light patches in the mysterious underwater world.
https://youtu.be/C-WQ8Dd44Xw

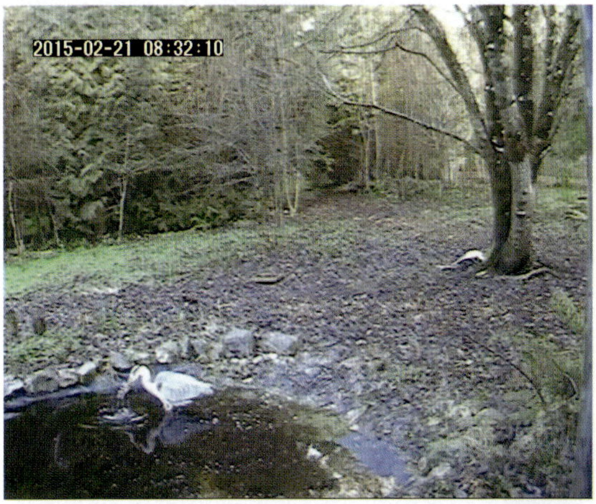

Figure 8.29 Heron frog catcher.
https://youtu.be/ul8Tf-eTTqA

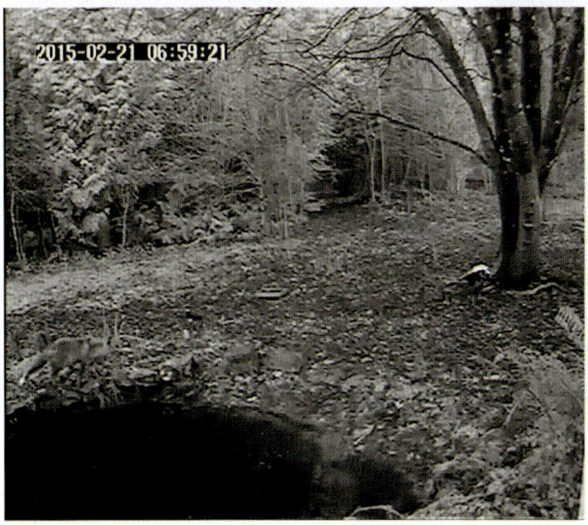

Figure 8.30 The heron is disturbed by a fox.
https://youtu.be/DW-zLZolkPg

The tawny owl in Figure 8.31 arrives in a flurry, picks up a frog and flies off. This video is not of a high quality as it was very dark and the camera too far from the owl, but it shows interesting behaviour, so has been included.

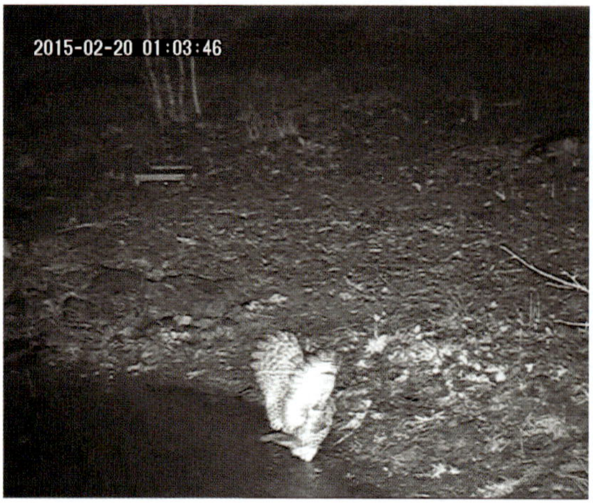

Figure 8.31 Tawny owl frog catcher.
https://youtu.be/3LNERsfi6Kc

Rock pools

The surface of the rock pool is still and reflects the pool contents. Tiny fish in Figure 8.32 can be seen to swim about amongst the seaweed which has many colours and shapes.

The limpets and sea snails in Figure 8.33 move very slowly as they graze the surface of the rocks for food, but if the video is speeded up, their constant motion is easy to sea.

Movement of the surface of the rock pool in Figure 8.34 causes patches of light to flash across the pond floor. It requires close watching, but a rock pool predator fish lies in wait for a tiny shrimp, pounces and seems to miss. Then another fish leaps into action.

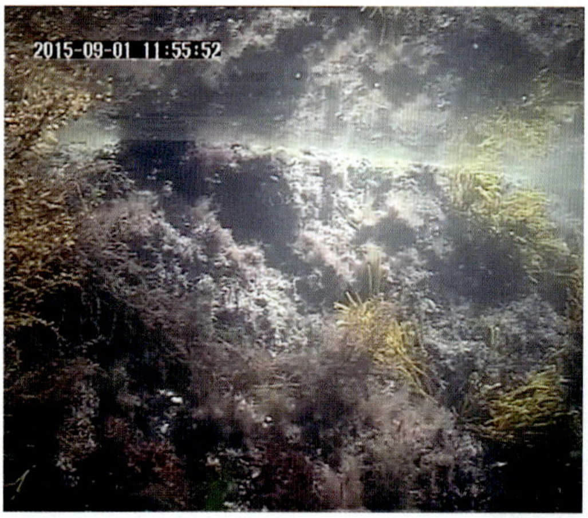

Figure 8.32 Rock pool with tiny fish and surface reflections.
https://youtu.be/HVHqrQ7NYN8

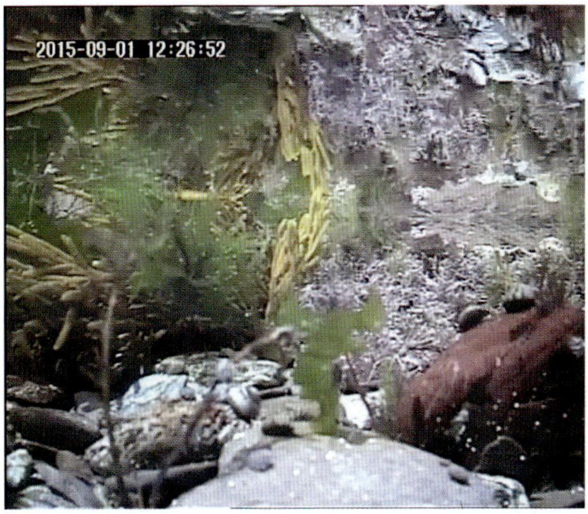

Figure 8.33 Limpets and sea snails move slowly, but are in constant motion.
https://youtu.be/O_XCc3i6R9g

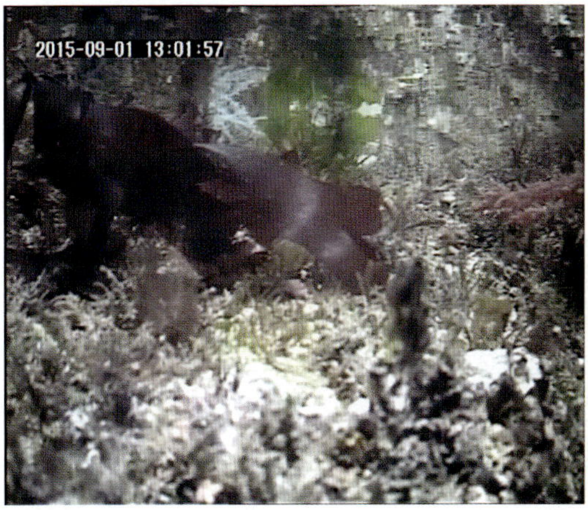

Figure 8.34 Predatory fish try to catch a shrimp.
https://youtu.be/DVZxoM5Wm24

Badgers

A sleepy female badger in Figure 8.35 leans her head on last year's cub and nearly dozes off, before the need for a scratch wakes her up and she starts foraging with her cub.

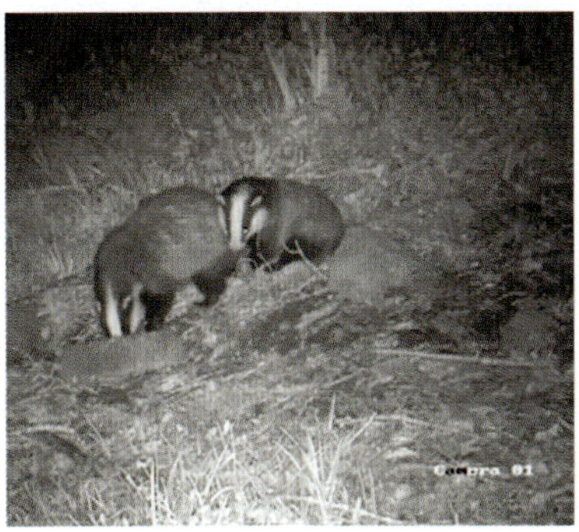

Figure 8.35 A sleepy female badger and her cub.
https://youtu.be/BVpalb5nKh4

Bird bath

A natural bird bath filmed with an HD-TVI camera in shows more than birds as seen in the video shown in Figure 8.36. The water is used for drinking by a shrew and a mouse.

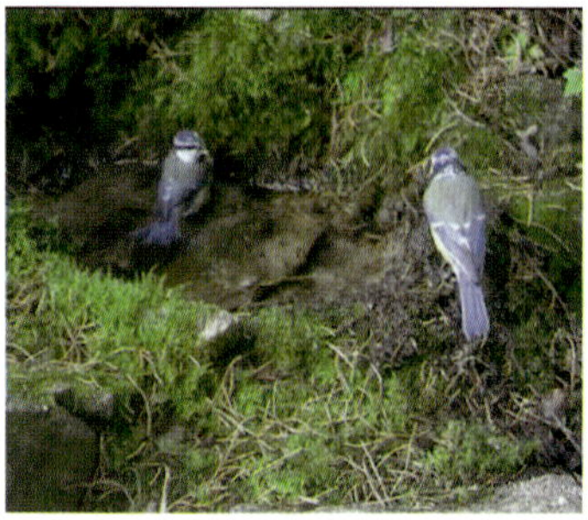

Figure 8.36 Bird bath action.
https://youtu.be/GZNvRe4mGac

Owls

The video described by Figure 8.37 shows a pair of tawny owls that meet on a branch, then start courting by preening each other. This was filmed when it was getting dark, but not quite dark enough for the camera LEDs to switch on, making the video a little grainy.

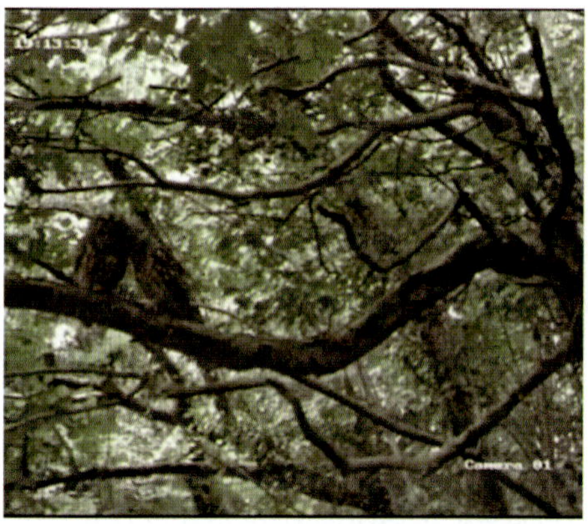

Figure 8.37 Two tawny owls courting on a tree branch.
https://youtu.be/wg2SUh-ohb8

9. Scientific case studies

These case studies are based on pilot research with Natural England and the Woodland Trust.

- 9.1 Fish monitoring.
- 9.2 Monitoring bats in woodland.

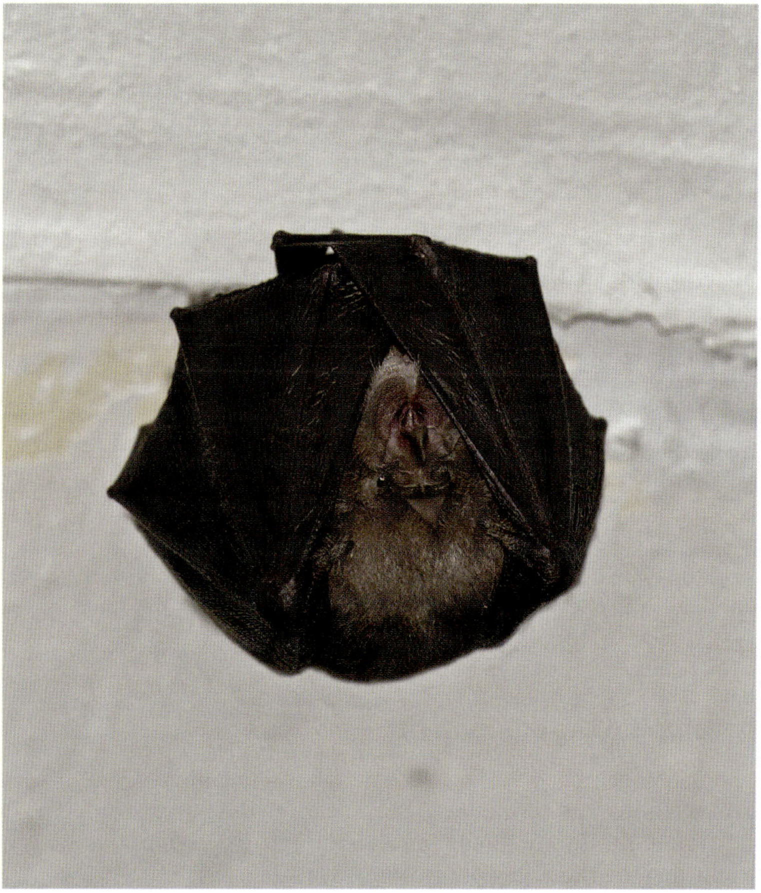

Figure 9.1 Lesser horseshoe bat (*Rhinolophus hipposideros*).

9.1 Fish monitoring using a submersible camera – a pilot study

This study was carried out using a portable CCTV system that I designed and developed for fish monitoring as a project of interest to Natural England in the Bovey Valley Dartmoor Natural Nature Reserves.

9.1.1 Introduction

In 2008, the Environment Agency (Washburn *et al.* 2008) devised a video system for counting fish, designed for the precise environment of the fish pass and at an estimated cost in 2008 of £5,000. Fish passes with counters tend to be in wide, deep rivers, often of 20 m width or more. The video system is frequently used in conjunction with a resistivity fish counter (Eltringham fishings on the Tyne 2015).

Electrofishing is the system used at present in small to medium rivers which can be waded to an average of thigh depth, and gives accurate counts of fish. With appropriate expert surveying of the stunned fish, information can be gleaned on species, size and weight. However, as the fish are stunned by the process, monitoring is not possible and little behavioural information can be gained. There is also evidence that a proportion of fish can be physically damaged by the process, commonly 1–5% but occasionally up to 50% (Beaumont *et al.* 2002).

My submersible camera and portable CCTV system has been developed and piloted to explore the more natural environment of rural rivers, with the aim of simplifying equipment and reducing cost and complexity. Because the fish are not disturbed, behaviour can be monitored and habitat preference can be established.

9.1.2 Aims of the project

1. To develop the best equipment for use.
2. To determine the conditions that give the clearest image.
   (a) Effect of weather and light conditions.
   (b) Water clarity.
   (c) River turbulence.
   (d) River depth and linear speed.
3. To determine the field of view of the camera.
4. To determine how fish size can be estimated.
5. To list shortcomings of the method and suggest ways to overcome these.
6. To carry out a short trial to monitor fish under the best conditions.
7. Suggest research/monitoring that could use this method.

9.1.3 Developing the equipment

There are very few submersible CCTV cameras available, and initially an adventure camera such as GoPro was considered. That idea was quickly discarded as the battery needed recharging after a few hours, and recording was continuous with no motion detection recording or scheduling of recording available.

After consideration, a system shown in outline form in Figure 9.2 was used.

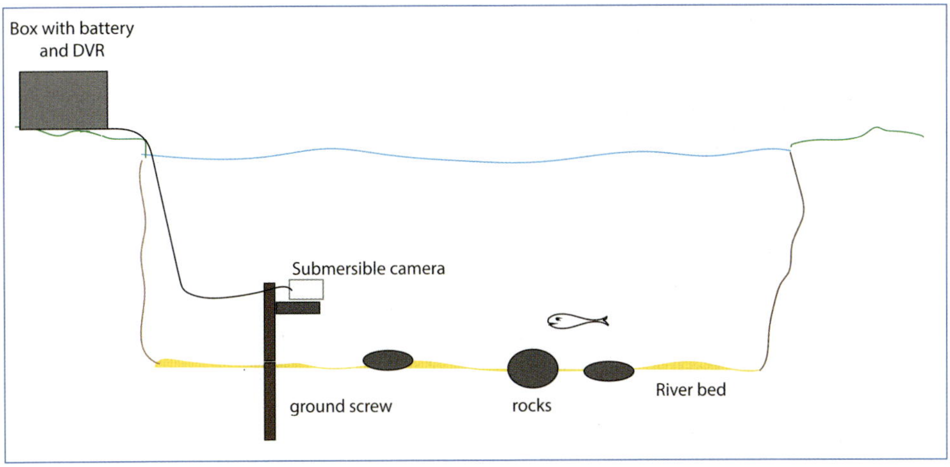

Figure 9.2 Fish study setup.

9.1.3.1 Camera

The camera used was a submersible bullet camera with infrared LEDs which give illumination when dark (although this was not used).

- Lens focal length 3.6 mm.
- Camera resolution 500 TVL.
- Voltage 12 V DC.
- Current 120 mA.
- Cable length 25 m.
- Connections on the cable were power (2.1 mm barrel plug) and BNC for video converted to male RCA with an adapter.

9.1.3.2 Camera support

A rig was designed to support the camera underwater, and allow some adjustment of position (Figure 8.3). The initial setup tied the rig to a wooden post, but this caused problems as it was not always easy to fix the post into the river bed where it was rocky. In addition, carrying a post and heavy hammer as well as the camera equipment and waders made it difficult to travel on foot. Hammering in the posts prior to testing was considered, but this reduced flexibility in experimentation, and also made it clear to potential thieves where the equipment was situated. As cameras had been stolen from the area earlier in the year, and this was a popular walking area, the possibility of theft was a major consideration.

The final design consisted of an adapted rig which fitted over a ground screw as shown in Figure 8.5 and Figure 9.3. A 500 cm ground screw could be screwed into the bed of the river relatively easily using the handle of a large screwdriver as a turning handle. Once the ground screw was in position, the camera and the adapted rig were fitted over the ground screw.

The ground screw and rig were relatively lightweight and could easily be carried in a rucksack with the other equipment. Other advantages were that the screw/camera

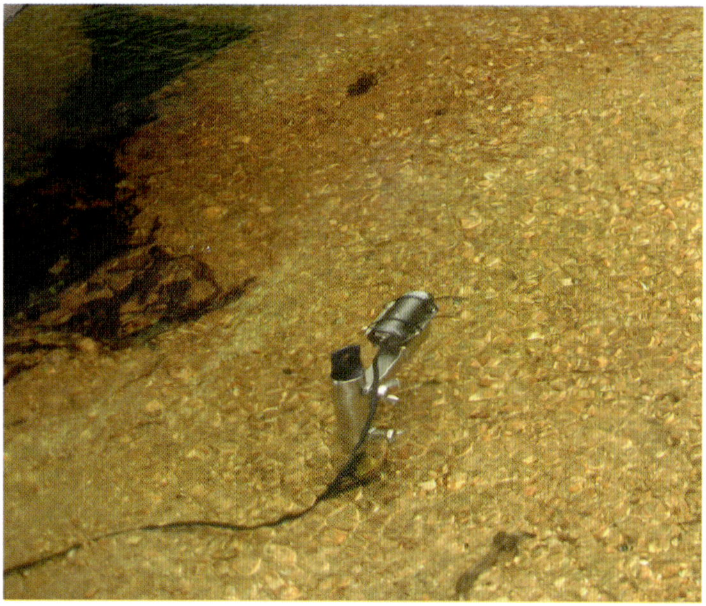

Figure 9.3 Camera and rig mounted on a ground screw. Both are several centimetres below the water surface, although this is difficult to see due to the clarity of the water.

assembly was below the water surface, making it less liable to be noticed and hence stolen. The assembly was later painted black for camouflage and stones placed round the base of the screw so that the assembly looked like a rock, in the hope that fish behaviour would not be affected by its presence.

9.1.3.3 Battery power

12 V DC was required by the camera and by the recorder. A 22 Ah lithium polymer battery was used. This battery lasted 3 days before charging was required.

Higher-capacity lithium batteries are available but at much higher cost.

Also tested was a 110 Ah lead-acid battery. This was heavy (22 kg) and it was considered that this would only be used where a solar panel could be set up to keep the battery charged, as this would reduce the need to move the battery back and forth for charging. A 30 W solar panel with a charge controller was tested over a period of several weeks, using the setup in Figure 9.4. Although several days had no sun, the solar panel kept the lead acid battery charged when current drawn by recorder and camera was 500 mA or less (as here). For higher currents, a 60 W solar panel would be required.

9.1.3.4 Recorder

In order that current draw was a low as possible, a portable DVR was required. The Genie CCTV SD-DVR was suitable. The important parameters were:

- Current draw 150 mA.
- Motion-detection recording with variable sensitivity.

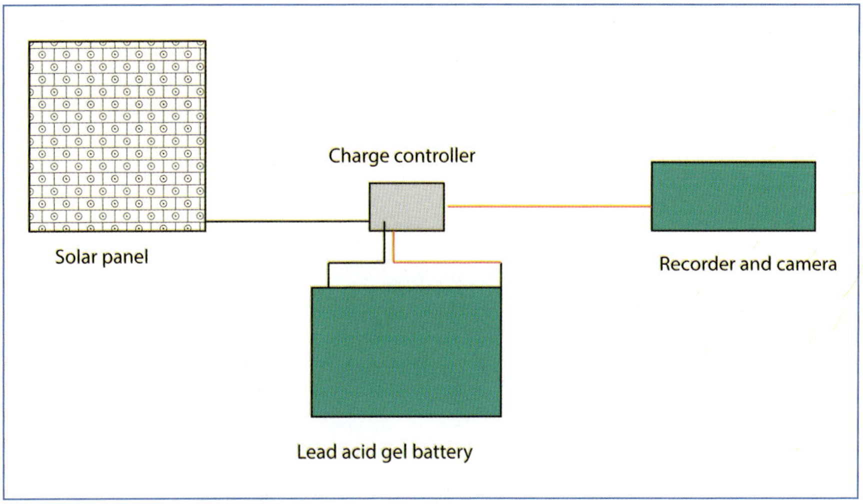

Figure 9.4 Solar panel setup.

- Pre-record of about 10 s.
- Scheduled recording possible together with motion detection.
- Post-event recording time adjustable.
- Small and lightweight.
- Uses a 32 GB SD card.
- Has its own software for pre-selection and pre-cutting of videos.

'Ruggedised' DVRs are available and are more suitable for outdoor conditions, but these cost 5–10 times as much as the SD-SVR.

9.1.3.5 Weather protection

As neither batteries nor DVR were weatherproof, they had to be kept dry. This was achieved with a double-case system. Recorder and battery were housed in a Peli case, with connections for the video signal from the camera and power to the camera via holes drilled in the side of the case. Waterproof connections were fixed to the holes. Due to the possibility of the recorder overheating in summer, the Peli case lid was lowered but not closed tightly, so that there was a circulation of air. This was not necessary in cold weather. The whole assembly, together with surplus camera cable, was placed in a waterproof black plastic box. The box had a hole drilled in the side to admit the camera cable and assist ventilation in summer.

9.1.3.6 Theft protection

Theft protection was necessary when the equipment was left on its own for tests lasting more than a few hours. The large plastic box had a chain wrapped round it and the chain was in turn wrapped round a tree or rock and padlocked into place. Camouflage material was placed on the box lid and the box placed in undergrowth on a side of the river without public access.

9.1.4 Factors affecting image clarity

9.1.4.1 Effect of weather

When it was bright, the camera image was very clear as shown in Figure 9.5. It was possible to see clearly across the stream to the other bank.

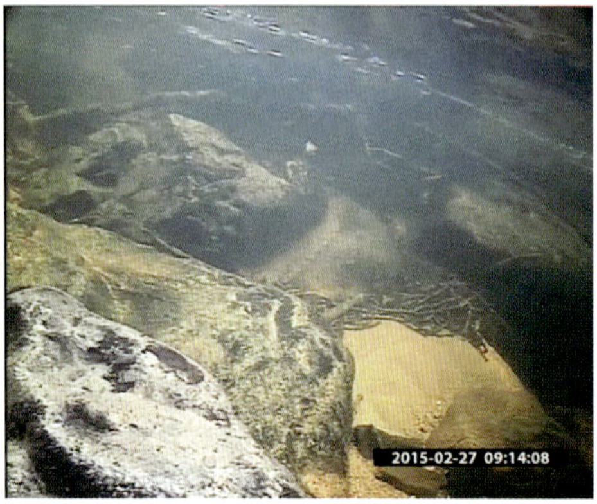

Figure 9.5 Image clarity in bright conditions.

When it was raining heavily, water clarity was affected as shown in Figure 9.6. The stream reacted quickly to increased water runoff from the land, and the resulting increased water flow created more turbulence and stirred up debris in the water.

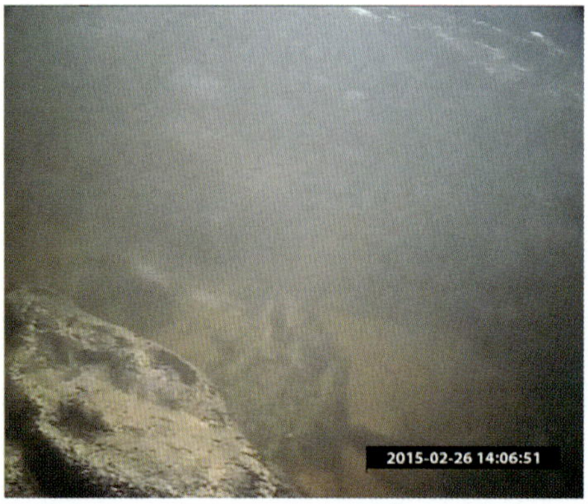

Figure 9.6 Lack of image clarity in rain.

9.1.4.2 Effect of light

When it was dark, the image became black and white and detail was hard to see (see Figure 9.7).

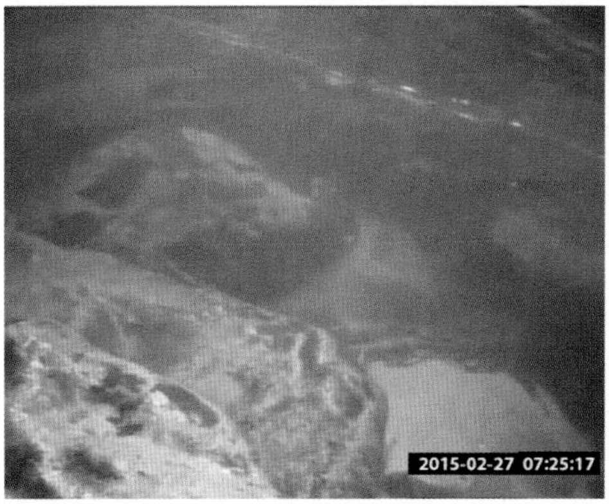

Figure 9.7 Image becomes black and white if light levels fall.

When the LED lights came on (see Figure 9.8) the light reflected off the passing debris, making it impossible to see anything. This might not be a problem in ponds where the water is still.

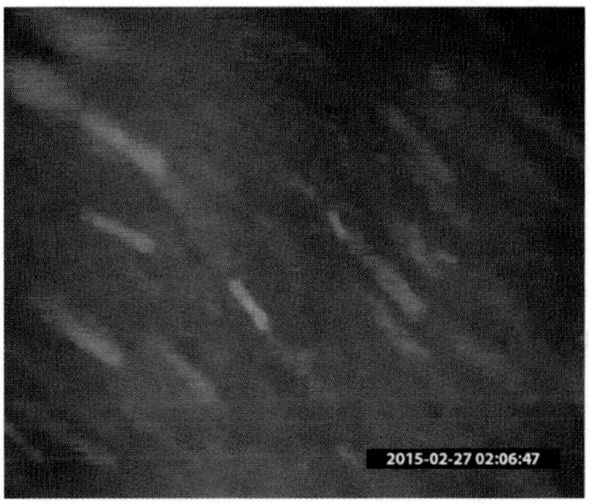

Figure 9.8 The image becomes obscure at night.

It was concluded that there would be no point in filming at night, and the DVR was programmed for motion detection in daylight only.

9.1.4.3 Patches of light

Further experiments showed that bright sunlight falling on patches of the river bed lit up the scene as shown in Figure 9.9. This was an advantage when these patches were stationary. However, when they were moving, the patches of light were regarded as motion by the recorder, giving rise to 'empty' videos.

It was concluded that the ideal lighting condition was bright cloud, as this gave even, bright illumination without moving patches of light on the river bed.

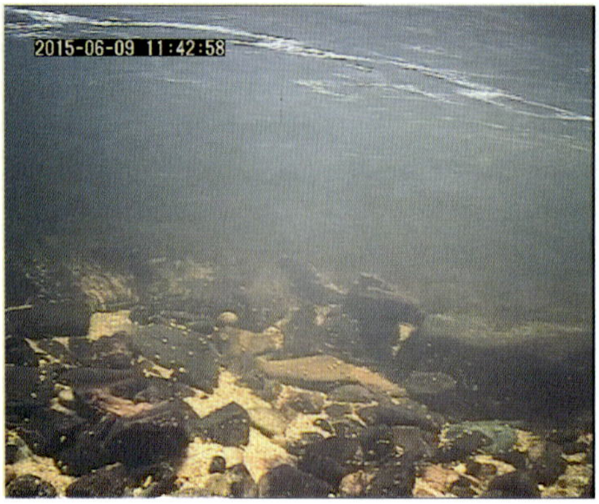

Figure 9.9 Bright sun lights up the scene but can also trigger motion detection recording.

9.1.4.4 Water clarity

Water clarity is the most obvious requirement for image clarity. In the Bovey Valley, the River Bovey is very clear, and the water itself gives ideal conditions for image clarity. However, it is a fast-flowing river, especially after heavy rain, and a large amount of debris is carried by it. Small leaves, twigs and other matter continually flow past, especially near the surface. This can affect water clarity, especially in autumn.

9.1.4.5 River turbulence

In many places the river is turbulent as it passes over and between rocks. Air bubbles and other material become suspended in the water, greatly reducing image clarity. Calm water with a smooth flow, or deep pools with minimal movement, were the best conditions for water clarity.

9.1.4.6 River depth and linear speed

Assuming that the volume flow of the river was approximately constant along the length of the section of the Bovey Valley for the period studied (there being no major tributaries), it was estimated that the linear flow would depend on the depth and width of the river at any point. At places where the river was wide or deep, the linear flow would be less, and thus turbulence less, than a narrow or shallow patch of water, especially if there were few rocks.

Ideal flow

In the section shown in Figure 9.10, the river bed was flat and the river wide and straight (some apparent man-made design was observed), and there were few stones on the river bed. The water depth was approximately 60 cm, and linear flow was approximately 0.25 m/s. Water clarity was exceptional due to the lack of turbulence and the slow linear flow.

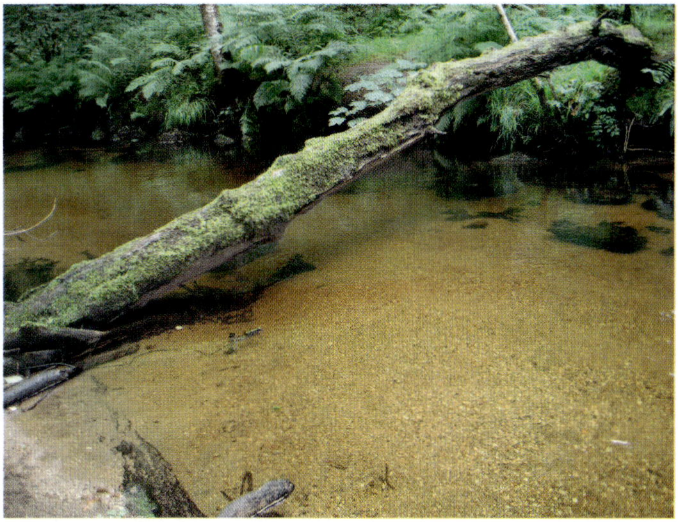

Figure 9.10 Ideal calm river flow.

9.1.4.7 Underwater view

The lack of features in the foreground in Figure 9.11 made estimates of distance difficult, although fish are easier to see against sand. Distance from the camera to the front of the large foreground rock, measured with a tape, was 1.75 m. Distance to the back of the rock was 2.75 m. Distance to river bank (just visible) was 5.5 m.

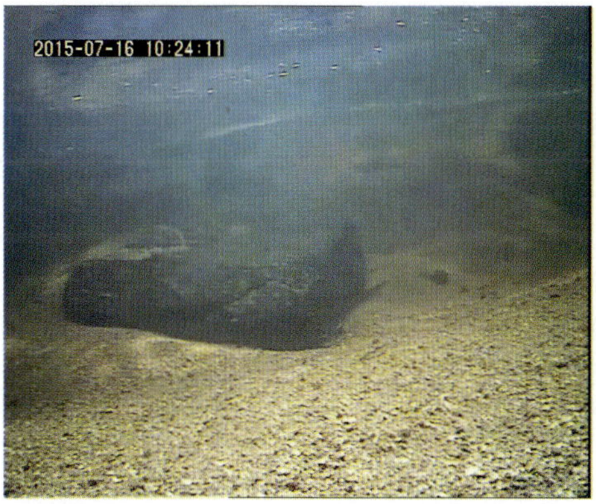

Figure 9.11 Underwater view from the camera.

Deep water

A pool upstream from the river section shown in Figure 9.10 was about 1.5 m deep. Image clarity was very good, especially as the camera could be placed below the surface layer where most debris and air bubbles were carried. Stones on the river bed made it easier to judge distance and size. There was sufficient sand to make fish easy to see. The fish tended to stay in the bottom half of the pool (see Figure 9.12).

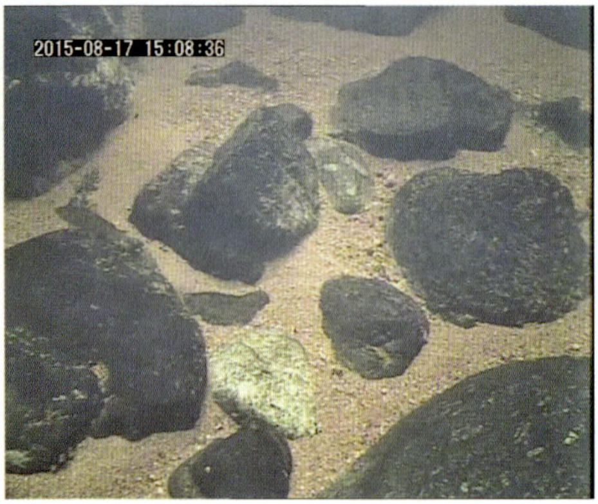

Figure 9.12 Perfect image in deep water.

Conclusions

From these observations the ideal conditions for initial fish monitoring test were judged to be:

- Clear water.
- Bright cloud above.
- A deep pool with minimum turbulence.
- Rocks and sand evenly distributed on the river bed.
- Daylight.
- Even, slow river flow.

9.1.5 Field of view

The field of view of the camera had to be established, to allow the estimation of the area of river being monitored at any time.

A plastic tape measure was placed along the flat bottom of a shallow pond and weighted down so that it was straight. The submersible camera was placed at the start of the tape and was connected to a battery and recorder so that the measurement process would be recorded on video.

A piece of wood of known width was placed underwater across the tape at 10 cm intervals from the camera for a total distance of 130 cm, which was the maximum distance that the pond base was flat.

The actual size of the viewing frame was estimated from the resulting video for each distance from the camera, by comparing the actual width of the wood with the apparent width in the video. For example, if a 25 cm wide piece of wood placed 40 cm from the camera took up half of the video frame, the horizontal field of view at 40 cm from the camera would be twice the width of the piece of wood. Thus the field of view would be 50 cm at that position.

The field of view was measured for each distance the piece of wood was placed from the camera. The measurements were repeated and average values were used to plot Figure 9.13.

The trend line was calculated and thus the horizontal field of view at greater distances could be estimated.

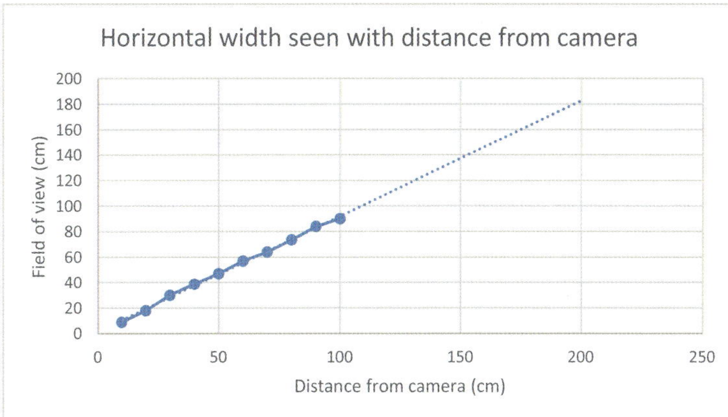

Figure 9.13 Field of view underwater.

From the graph, a field of view diagram can be drawn. At a distance of 2 m from the camera, the width covered by the camera was approximately 1.8 m. At a distance of 4 m from the camera, the width covered by the camera was approximately 3.6 m as shown in Figure 9.14.

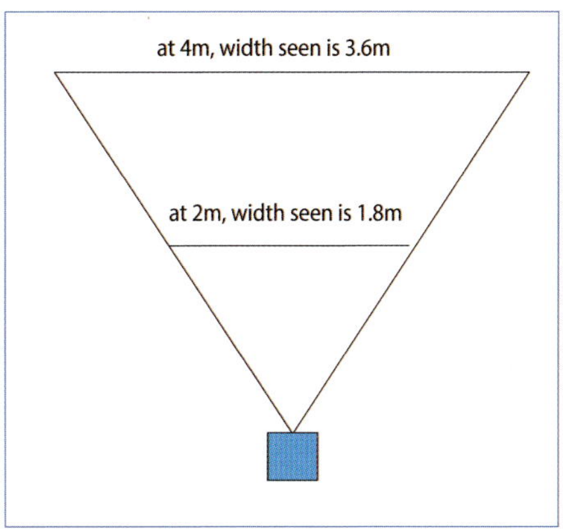

Figure 9.14 Fishcam field of view.

The vertical field of view was estimated to be 0.8 of the horizontal field of view as a result of the sensor height being 0.8 of sensor width. Thus height seen at 2 m would be 1.4 m and at 4 m would be 2.9 m (the method is not accurate to two decimal places so the figures are rounded to one decimal place).

Field of view can be calculated from the camera's stated angle of view in air using simple trigonometry, but the field of view in water is much less than in air due to the refracting effect of water. In this case the angle of view in water was approximately 48°.

9.1.6 Determination of fish size

9.1.6.1 Perspective distortion

The camera used had a relatively wide angle of view, and it was necessary to establish how the apparent size of underwater objects varied with distance from the camera, as wide-angle lenses are known to cause noticeable perspective distortion depending on the object distance from the camera – objects close to the camera look abnormally large and distant objects look abnormally small.

Method

A plastic tape measure was placed along the flat bottom of a shallow pond and weighted down so that it was straight. The submersible camera was placed at the start of the tape and was connected to a battery and recorder so that the measurement process would be recorded on video.

A 25 cm wooden 'fish' was placed 10 cm from the camera, left for a few seconds then moved to a position on the tape 20 cm from the camera. The process was repeated at 10 cm intervals for a total distance from the camera of 100 cm.

The measurement process was repeated with a 40 cm 'fish'.

A range of stills was taken from the videos of the measurement process for each 10 cm measurement. Each still was viewed on a monitor in Photoshop and the apparent fish size measured from the rulers available in the software. It should be emphasised that these were comparative measurements only.

From Figure 9.15 it can be seen that apparent 'fish' size increases at a distance of 50 cm from the camera and increases rapidly as the distance from the camera decreases. At distances greater than 50 cm from the camera, the apparent size falls off very slowly. For cameras with a smaller angle of view, the size distortion would be less.

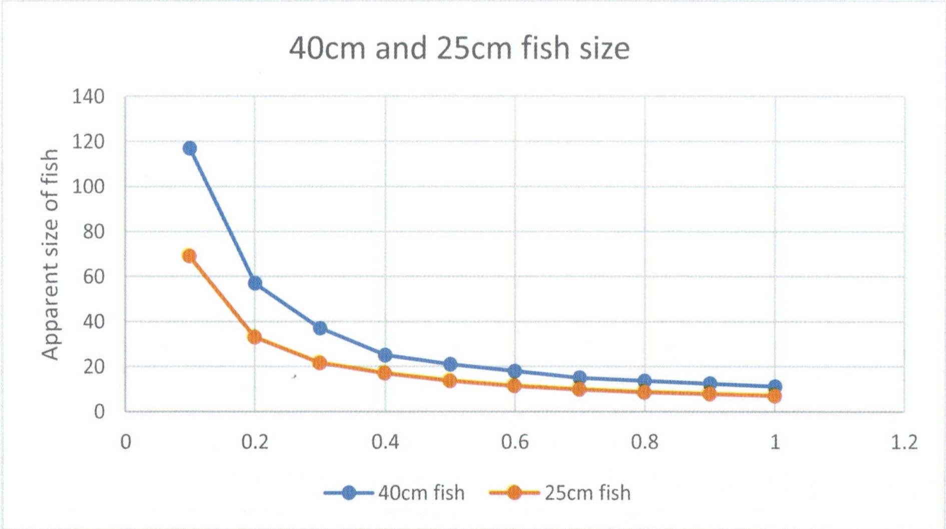

Figure 9.15 Perspective distortion. Apparent fish size varies in a non-linear fashion with distance from the camera.

Conclusions

This experiment shows that perception of the size of a fish in a video can be misleading as perspective distortion makes size variation from the camera non-linear, especially within 50 cm of the camera.

9.1.6.2 *Estimating fish size*

If the accurate measurement of fish size is regarded as important, markers could be placed on the river bed at known distances and the distance of the fish from the camera estimated from its position relative to the markers. The actual size of the fish could then be determined from the field of view graph.

For example:

- If a fish is estimated to be at a distance of 100 ± 10 cm from the camera, the field of view is 90 cm, taken from the trend line of the field of view graph.
- If the fish takes up 0.25 of the width of the video, its size is 0.25 × 90 = 22.5 cm.
- Taking errors into account, actual size is quoted as 22 ± 3 cm.

Alternatively, fish size could be estimated by mapping the riverbed roughly, measuring approximate size and position of rocks on the river bed. In clear water, this can be facilitated by taking an overhead photograph.

For example:

- A rock is 40 ± 2 cm wide and a fish takes up 0.25 of its length.
- The width of the fish is 0.25 × 40 cm = 10 ± 1 cm.

9.1.7 General issues to consider

The distance that fish can be seen clearly by the camera is a minimum of 2 m, and a maximum of 5 m in ideal conditions. This could mean that wider rivers would have to be netted at positions where monitoring is to take place so that fish would be constrained to a narrower channel. This technique is used in electrofishing. However, I believe it would be easier to select areas that suit the equipment limitations, rather than try to adjust the environment to suit the equipment.

There are limits to the conditions in which the fish can be monitored, as turbulent areas and rivers with poor clarity cannot be used. However, there is also a great variety of conditions to be found along the length of any river and, with careful study, suitable monitoring sites will be found.

If the river is relatively narrow, the camera could be pointed downstream. This allows a greater distance to be monitored and also reduces the amount of debris impinging on the lens, which in turn reduces unwanted motion-detection recording. It also means that the lens will require cleaning more often as water is not flowing into or across the lens to clean algae from it.

9.1.8 Video analysis

Under ideal conditions, motion-detection recording at low sensitivity would be triggered only by fish moving past. These are the conditions used in the Environment Agency fish counter at Riding Mill on the River Tyne for example, where motion-detection video recording is triggered when a fish is detected by the electric fish counter. In this case, fish move through a fish pass, which gives ideal conditions for video.

In the River Bovey, conditions are natural and the constant flow of debris carried by this and other fast-flowing rural rivers can produce a large number of videos in less than ideal conditions.

For example in an autumn trial where there were a large number of leaves carried by the river, a 3 day recording period produced 10 hours of video. However, by using fast forward, videos with no interest were quickly discarded, leaving those of interest for more detailed analysis. Two hours were spent processing the videos.

Before carrying out any trial, the conditions and equipment should be tested to determine the likely number of videos and monitoring projects designed accordingly.

9.1.9 A selection of trial results

A number of trials were carried out under different conditions and three contrasting videos were produced:

Video 1

Video 1 (see Figure 9.16) shows small fish in Becka Brook, a tributary to the River Bovey. The fish in this and the other videos recorded at the same place stayed in a small area for several days. The fish invariably pointed directly upstream when not looking for food. Becka Brook was about 2 m wide at this position and was very fast flowing. The stream was about 40–50 cm deep. Because of the rocks in the stream, the camera could not be placed ideally and the water flowing at the surface was present in the videos, causing frequent motion-detection recording. The larger fish was estimated to be 10 ± 1 cm by comparing its size with the measured size of the adjacent rock.

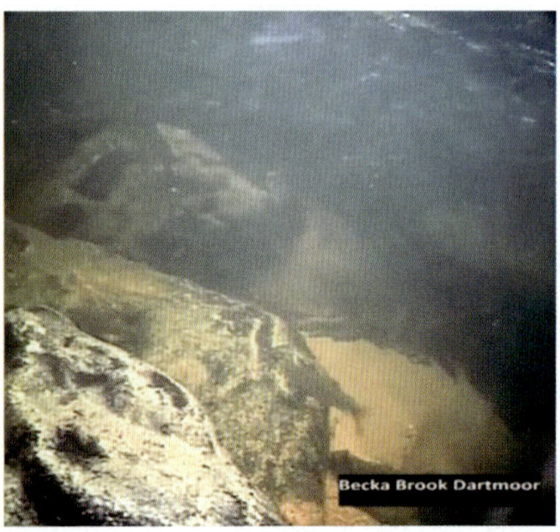

Figure 9.16 Two fish in Becka Brook.
https://youtu.be/qWYZ3ojdhKQ

Video 2

Video 2 (see Figure 9.17) was taken with the camera rig handheld in a deep pool with ideal conditions. With the rig fixed, these conditions would give a much lower number of videos triggered by non-fish events.

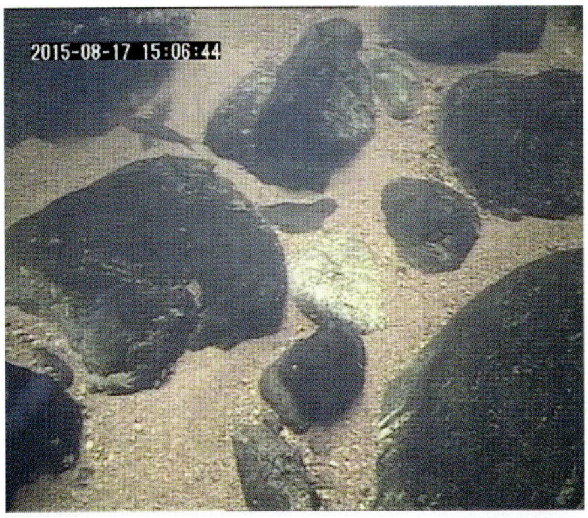

Figure 9.17 Fish in a deep pool.
https://youtu.be/oEok3ESPCcw

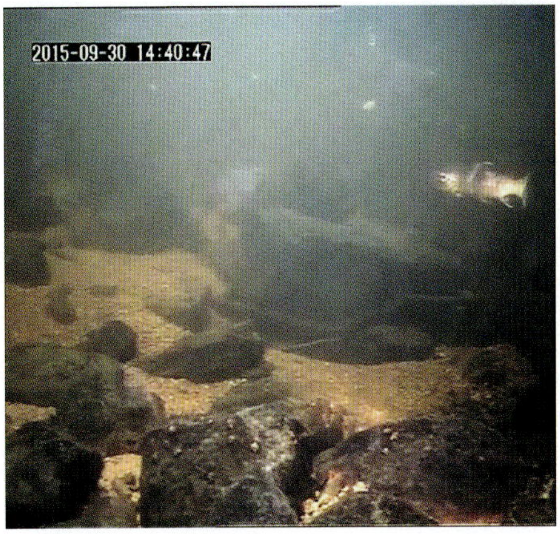

Figure 9.18 A larger fish in a deep pool.
https://youtu.be/GaHDKdbO508

Video 3

Video 3 (see Figure 9.18), taken at the start of autumn, was filmed in a deep pool in the River Bovey. The aim of filming in this deep pool was to determine if the depth of the pool would attract larger fish, which indeed it did. This fish was present in the same small area of the river under a bank for the 2 weeks of the trial. As with the fish in Video 1, it stayed pointing upstream except when looking for food. The fish was estimated to be 22 cm × 3 cm from field of view measurement.

Several smaller fish appeared some days after this video, and each stayed in its own section of the river where it was shallower. The videos produced later in the trial were not very clear, partly because of the autumnal debris flowing downstream, partly because it was quite dark under the trees, but mainly because it was discovered that the lens of the camera had a film of algae on it.

9.1.10 Suggestions for further study

The study raised many questions about fish behaviour and opened up many possible areas for further study:

- Monitoring fish behaviour.
- Do fish tend to inhabit the same small area as suggested by these trials?
- Determination of the favoured habitats.

- Is the size/type of fish present dependent on linear speed, turbulence, clarity, habitat (bed of river, presence of plant growth), shade or light in a particular location?
- At what depth do fish tend to be most of the time and does this depend on fish size?

9.1.11 Conclusions of the case study

There are many possibilities for using submersible cameras to study underwater life, if care is taken to be aware of the conditions giving the clearest videos. Due to the large number of videos that can result in less than ideal conditions, monitoring would be best carried out for several days at a time rather than for long unbroken periods.

In the small number of trials carried out for this pilot study, the majority of fish seemed to stay in a small area for days at a time, some for much longer, which also suggests that short recording periods of several days may be a suitable monitoring period.

Use of a submersible camera gives information about fish behaviour and fish presence in different habitats. By adapting the camera rig, deep pools can be sampled, and in fact deeper water can give rise to clearer videos. Deep pools are not usually sampled by electrofishing, so this method could provide valuable information.

The field of view of the camera is small, so for a comprehensive survey, the camera should be moved to several sites near each other and the results combined. Alternatively several cameras could be used at once. In ideal viewing conditions, pointing the camera downstream would sample a larger area.

The cost of camera, battery and DVR was around £600. The equipment can be used for many monitoring subjects as described in this book, and is not limited to this type of study.

Although the setup was used to monitor fish, a similar setup could be used for monitoring aquatic mammals such as beavers.

9.2 Monitoring bats in woodland – a pilot study

It must be emphasised that it is an offence to disturb a bat roost. If you wish to use a CCTV camera near a roost, first obtain a license from Natural England, as was done here. Only low-intensity infrared LED lighting on the camera, or as a separate illuminator, should be used.

9.2.1 Introduction

In early 2015 I had developed a pioneering portable CCTV monitoring system and was trialling this with various wildlife species of interest to Natural England and the Woodland Trust on Dartmoor. I was introduced to the Woodland Trust team studying barbastelle bats (*Barbastella barbastellus*) in the Bovey Valley Woodland as part of the 'Moor Than Meets the Eye' partnership and was very keen to take the opportunity to use the portable CCTV system, especially as, to my knowledge, very few videos of *B. barbastellus* had been obtained.

I was fortunate that the team had discovered a *B. barbastellus* maternity roost using radio-tracking. The roost was in an old oak tree and being about 3–5 metres off the ground was accessible to a CCTV camera without tree climbing being required.

At present, bat surveys mainly use radio-tracking, bat detectors and emergence surveys, techniques which can mean a great deal of work at night, often without the visual detection of bats. The aim of the pilot study was to show that CCTV could be a useful additional tool for bat surveying by reducing the amount of night work such as emergence surveys, and producing visual material.

Although the roost was vacated a few days before the camera was set up, it was decided to leave it in place as it is known that *B. barbastellus* switches roosts frequently (Russo *et al.* 2004), and it was considered likely that the bats would return at some point.

9.2.2 Equipment

9.2.2.1 Cameras

Three different cameras were tried to give different viewpoints and resolution.

- Bullet cameras waterproof to IP66.
- All with variable-focus lens of 2.8–12 mm.
- Two with 630 TVL resolution and one with 1,000 TVL resolution.
- Voltage 12 V DC.
- Current draw 250, 550 and 700 mA depending on the camera in use.
- Infrared illumination supplied by 40 infrared LEDs.
- Connections on the cable are power (2.1 mm barrel plug) and BNC for video signal.

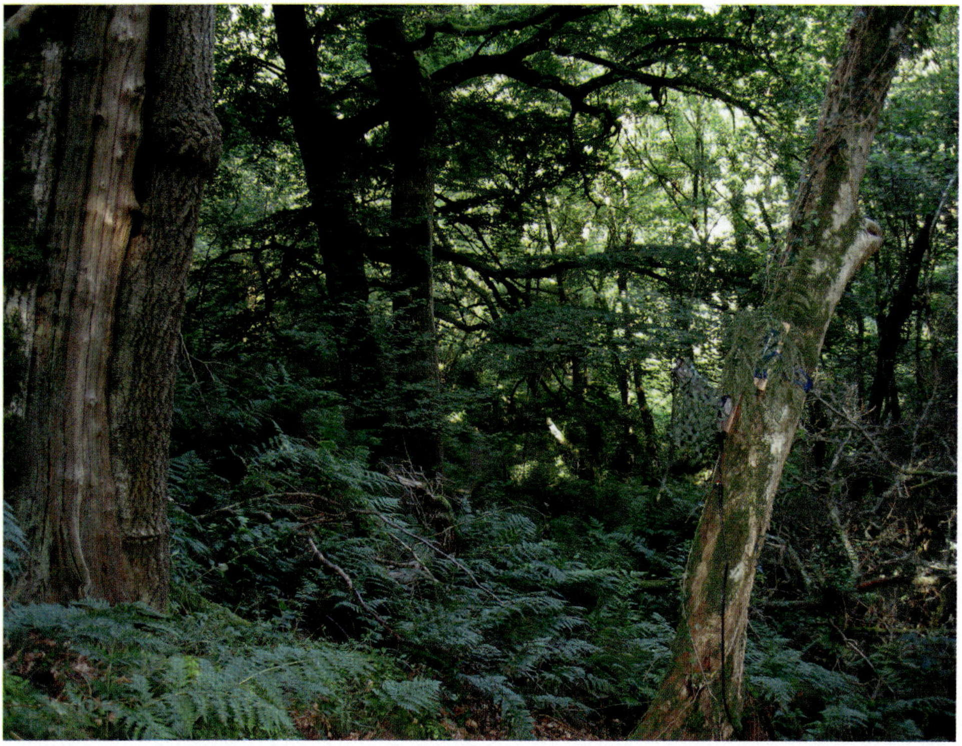

Figure 9.19 Camera and Wildlife Acoustics SM2 bat detector on the support tree on the right. The roost tree is on the left.

9.2.2.2 Camera support

Several methods of support were used so that the camera position could be changed to give differing views. The camera base was first fixed to a piece of outdoor plywood. This gave greater flexibility in positioning the camera as only two screws were required for fixing, rather than the four screws used for fixing the camera base itself. The plywood was fixed to a tree using two galvanised screws. Holes were drilled in the tree first to produce a smooth-edged round hole, to minimise bark damage.

The camera support had ivy wrapped round to camouflage it, and a piece of camouflage material was tied round the camera body with a cable tie. The image in Figure 9.19 shows the camouflaged camera on the support tree, with the cable and chain (which were still to be camouflaged). The box containing battery and recorder is to the right of the support tree.

To allow movement of the camera to different positions and further minimise tree damage, a ground spike, such as those used for fence posts, was hammered into the ground and a square wooden post (75 mm × 75 mm) placed into the ground spike and fixed using the bolts on the ground spike. Ground spikes with bolts are more suitable as this makes it easier to change the position of the post and also to remove the spike and post once the project has finished.

9.2.2.3 Battery

A 12 V DC 22 Ah lithium battery was used and changed every 2 days. As the project took place in woodland, using a lead-acid battery charged by solar panel was not considered as there was too much shading from trees. Later, a larger-capacity lithium battery was used which lasted 4–8 days, depending on which camera was used, with the 250 mA camera drawing the lowest current and hence drawing less from the battery.

9.2.2.4 Recorder

In order that current draw was a low as possible, a portable DVR was required. The Genie CCTV SD-DVR was chosen. The important parameters were:

- Current draw 150 mA.
- Motion-detection recording with variable sensitivity. Sensitivity at the medium level was used.
- Pre-record of about 10 s.
- Scheduled recording together with motion detection.
- Post-event recording time 10–20 s.
- 32 GB SD card used.

9.2.2.5 Weather and theft protection

For weather and theft protection, the recorder and battery were placed in a Peli case that had the lid lowered, but closed loosely, to allow for ventilation as it was summer. The Peli case was placed in a larger plastic box with a hole drilled in it for the cables to pass through. A chain was passed round the outer box, through the box handles, round a tree and padlocked. A square of EPDM pond liner was placed on top of the plastic box to prevent water from collecting on the lid, as this sometimes dripped inside the box when it was being opened, and could have damaged the electronics of the recorder. Camouflage material was placed over the plastic boxes and bracken placed on top.

9.2.2.6 *Bat recorder*

A Wildlife Acoustics SM2 bat recorder was fixed close to the roost tree to record bat calls. The aim was to use the sound recordings to confirm the species of any bats seen on video.

9.2.3 Method

After the camera had been fixed to the tree, the camera angle was adjusted so that the lens pointed towards the previously used egress points on the tree. Previous emergence studies made while the maternity roost was occupied had identified two main egress positions: to the bottom right and top left of the tree, as shown in the infrared image in Figure 9.20.

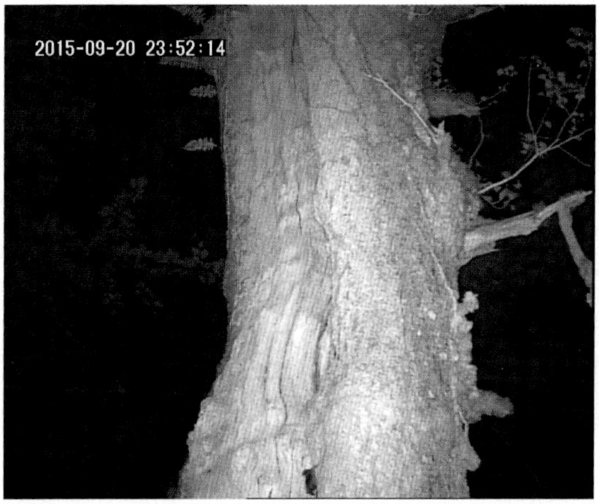

Figure 9.20 The roost tree egress points are bottom right and top left of the tree. A bat can be seen on the tree centre bottom.

The SM2 bat recorder and video recorder were time synchronised and the recording schedule set up for dusk to dawn recording from 21:00 to 24:00 and 0:00 to 06:00 every day. Two different schedules were required as the 'to' time must be later than the 'from' time for each day which means that a recording time from 21:00 to 06:00 would not have been accepted by the recorder.

Motion-recording sensitivity was set to normal (a middle value).

The post-record time was set to 20 s. This had the advantage that short events, such as a moth passing, would produce very short videos that would be easy to scan and discard. Any longer event, such as a bat flying about, would last as long as the motion, because the bat flying would continually trigger motion recording.

Moving foliage can cause long periods of recording producing videos of no value, so the camera was positioned to avoid moving vegetation. A little light pruning was undertaken of a few branch tips that protruded into the image and set off recording. Heavy rain is a major problem with CCTV as it can set off long recording periods especially as in this case the camera pointed upwards. Fortunately this was only a problem on two or three nights as tree cover helped to reduce the effect of rain causing motion-detection recording.

The software bundled with the Genie recorder made it easy to quickly run through videos and discard those of no interest.

9.2.4 Initial results

9.2.4.1 Summary

At the start of the project, the aim was to film bats on emergence, which required the roost to be in use. As it had been unexpectedly vacated prior to the study, it was expected that videos would not be obtained for some time, if at all. However, videos were produced after a few days. These videos were not, as expected, from bats using the tree as a day roost, but consisted of short night visits. The early videos showed one or two bats flying round the tree, landing on it, then flying off again, sometimes peering into the egress points. It appeared that the bats were checking out the roost, possibly keeping tabs on it for future reference, but some other behaviour was hard to explain.

Some bats flew onto and away from the tree several times in succession. In most of these cases there were no other bats present. The cause of this behaviour is unknown, although the hypothesis of those who have viewed the videos was that it could be due to juvenile bats improving flying skills or bats warming up flight muscles.

9.2.4.2 Analysis

There were three main types of activity:

- Bats flying in the vicinity of the roost with occasional inspections of one of the cavities in the tree.
- Bats landing on the tree, moving around on the tree or into a crevice for a short time.
- Repeated flying onto and off the tree trunk, sometimes four or five times in a short time span of 20 seconds or less.

Number of events

The videos recorded over the two months of monitoring showed that the number of short visits fluctuated, with days passing with no bat visits (as seen by the camera), interspersed with periods of more intensive visiting activity. This is shown in Figure 9.21. All visits were short-lived.

Number of bats per event

In most cases, a single bat was involved in each event, although without radio-tracking it was impossible to determine the total number of separate bats that visited the roost. Figure 9.22 shows that of the 96 separate events, 12 involved two bats and 3 involved three bats.

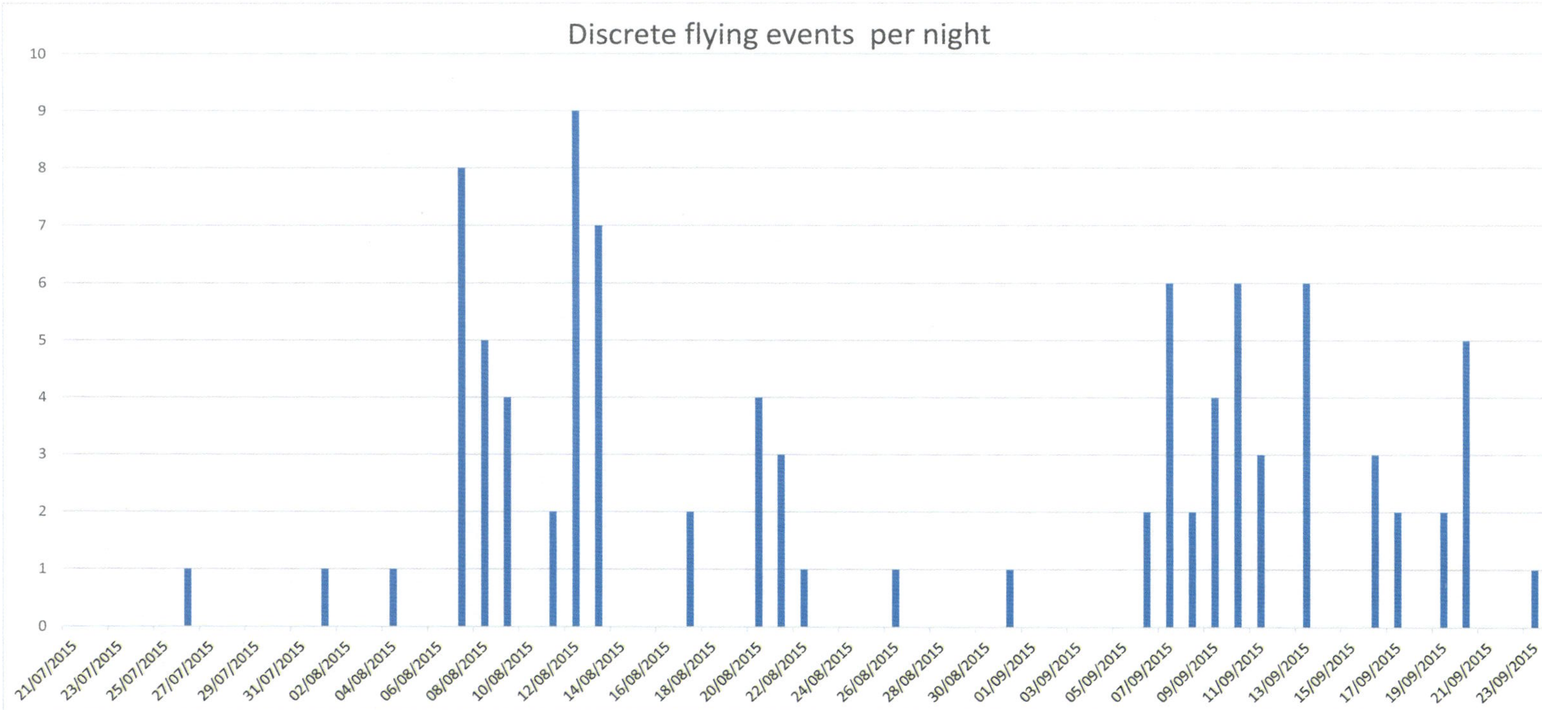

Figure 9.21 Discrete flying events and dates. A discrete flying event is defined as one more than 1 minute after a previous flight.

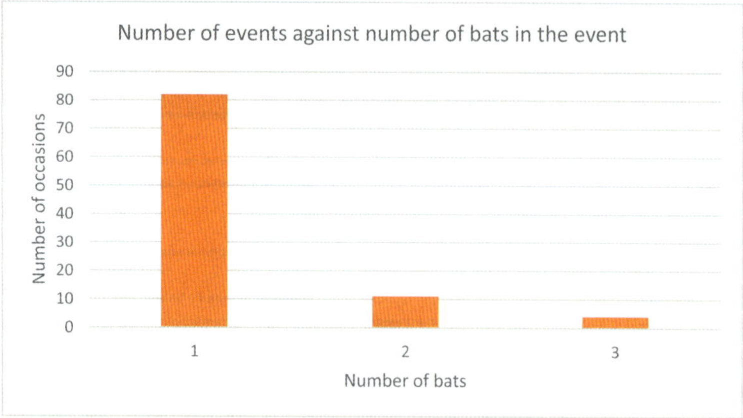

Figure 9.22 Number of flying events vs. the number of bats involved.

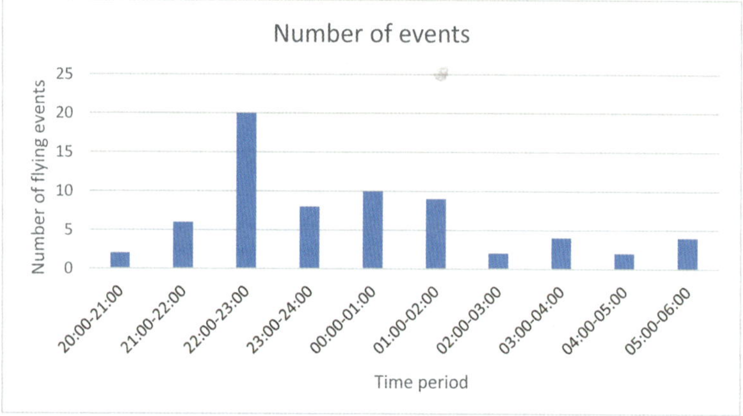

Figure 9.23 Number of events per time period.

When two or three bats were involved, there was some interaction between the bats on most occasions.

Timing of events

Initially recording was scheduled at dusk and dawn only, but as it quickly became apparent that there was no night emergence, timing was scheduled throughout the whole period from dusk to dawn. Figure 9.23 shows that flying events took place throughout the time period with a peak at 22:00–23:00.

9.2.4.3 *Bat species involved*

The time intervals of the flying events were compared with the bat calls on the SM2 bat detector for the same time interval.

In August, *B. barbastellus* registered 25 calls and 10 coincidental flying events. There were also a large number of flying events that did not have corresponding calls recorded.

Times did not coincide exactly with the SM2 calls in every case, but these were believed all to be *B. barbastellus* for four reasons:

- The only calls recorded at the time were *B. barbastellus*.
- In most cases the definite and suspected *B. barbastellus* flying events were very close in time.
- Where there was a discrepancy between time of recording and time of flying event, it was very short, between 5 and 30 s.
- Flying pattern and behaviour was very similar in all cases.

The discrepancy in times, and the fact that there were a large number of silent passes, led us to suspect that *B. barbastellus* may alter or omit echo-location in the vicinity of the roost tree. *B. barbastellus* is known to alter echo-location for different purposes (Selbert *et al*. 2015), and we hope to carry out more work next season on *B. barbastellus* echo-location close to a roost.

On other occasions there were 21 pipistrelle (*Pipistrellus* sp.) calls and calls from other bats, often at the same time, which made it difficult to determine which variety of bat was showing on the videos. *Myotis* sp. were also recorded.

For September, no *B. barbastellus* calls were recorded and long-eared bats (*Plecotus* sp.) were seen on several occasions, instantly recognisable by the length of ears and their hovering flight, although there were no calls recorded for them at the time of flight. In fact the most notable feature of the study was the fact that at least 30 flying events recorded on video had no associated calls detected at all.

9.2.4.4 Behaviour

There are different explanations mooted for bat behaviour in the vicinity of roosts, and much of the research on bats seems to be related to occupied day roosts and radio-tracking in foraging areas. There seems to be very little previous work similar to that described here, as the roost monitored did not appear to have been used as a day roost for the period of study. It had been a maternity roost for *B. barbastellus* and was vacated just before the study started, and other species of bat that had not occupied the roost were also recorded in the vicinity of the roost tree.

As the behaviour observed did not fall into the categories of night emergence, dawn swarming or foraging, any explanation without corroboration can only be speculation. As most flying events seemed to involve individual bats, it does not seem likely that social interaction or pre-mating behaviour could be involved. In most cases it appeared that the bats were checking out the roost, or stopping briefly, possibly to rest or to manage a large prey subject. The tree attracted a great deal of interest, although without a control of study of other random trees one could not draw firm conclusions as to how significant the interest in the roost tree was.

9.2.5 Conclusions

The aim of the case study was to pilot the usefulness of CCTV as an additional tool for the bat surveyor, and possibly reduce the amount of night emergence surveys normally necessary. The study greatly exceeded expectations and gave rise to some exciting results. Not only were many videos obtained, but what appears to be rarely seen behaviour was observed.

Many questions were raised by the study and further research is required, in particular to study behaviour at an occupied day roost and compare this with the behaviour at an unoccupied roost, as this one appeared to be.

The study suggests that not only can the use of CCTV reduce the need for manual emergence surveys, it also suggests that valuable research results could be obtained from the use of CCTV in the monitoring of bat behaviour in this way.

9.3 Videos

Bats

Two barbastelle bats appear to be copying each other's flight in the video from which Figure 9.24 was taken. This took place a week after the barbastelle bats in the maternity roost in the tree had switched to another roost. This could be a bonding pair or a juvenile and adult.

The barbastelle bat in Figure 9.25 appears to be checking out the roost that had just been vacated a few days before.

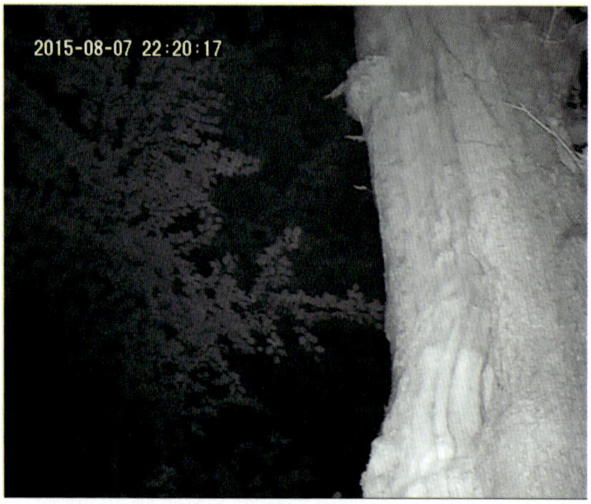

Figure 9.24 Two barbastelle bats flying in tandem.
https://youtu.be/6i4DUAAQ9T4

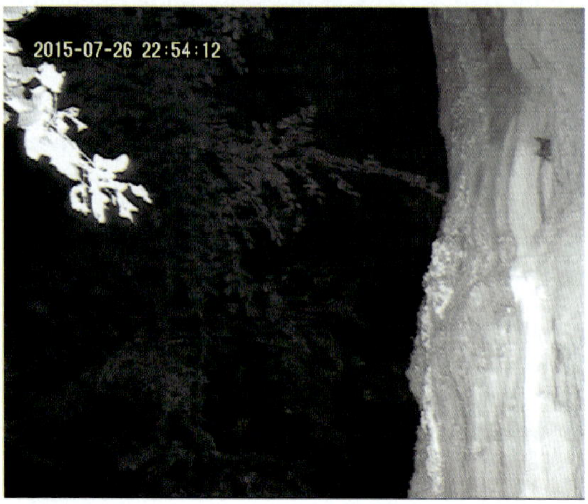

Figure 9.25 A barbastelle bat appears to be checking out a roost tree.
https://youtu.be/O_0whqdqCqw

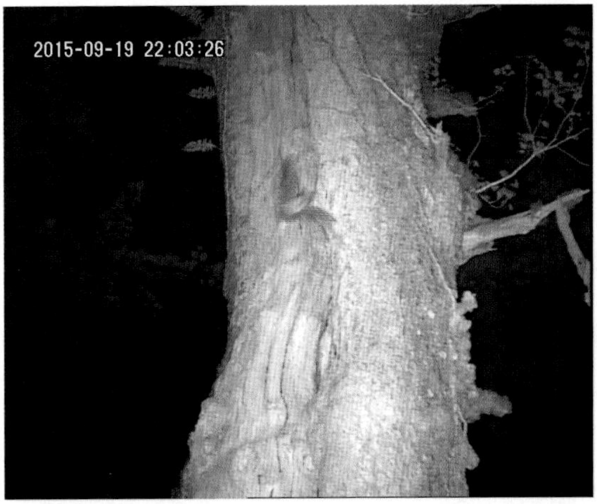

Figure 9.26 A long-eared bat hovers about and checks out the roost tree.
https://youtu.be/RRaAAMa4Qfg

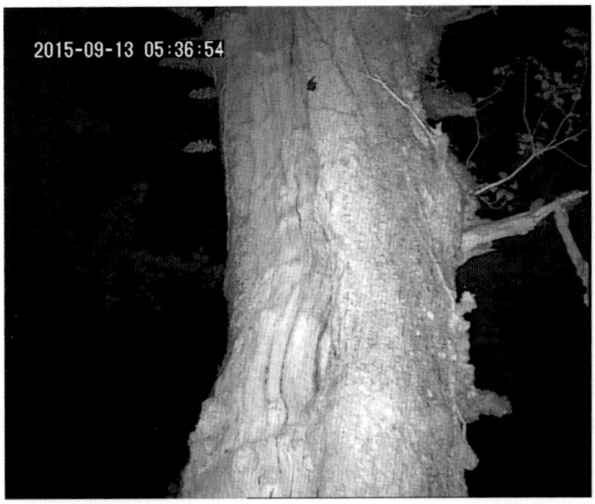

Figure 9.27 Bat repeatedly flying on and off the tree.
https://youtu.be/PTjwV9pBvqY

A long-eared bat (see Figure 9.26) is distinguished by its hovering flight and noticeable long ears.

A common behaviour seen was of a bat flying repeatedly onto the tree and then off again. This is shown in the video in Figure 9.27. There was speculation this was a juvenile improving flying techniques or just warming up flight muscles. There were no bat calls recorded at the time of flight to identify the bat species.

10. Suggestions for CCTV monitoring subjects

Several further areas for CCTV monitoring are suggested.

- 10.1 Birds.
- 10.2 Reptiles.
- 10.3 Insects.
- 10.4 Aquatic mammals.

10.1 Birds

Care should be taken that birds are not disturbed by any CCTV action, especially in the nesting season. The Wildlife and Countryside Act 1981 protects all birds, their nests and young from disturbance. Schedule 1 birds (listed in the Act) have special protection, and nests cannot be approached without a licence. A specific licence is required for photography. Schedule 1 birds include kingfishers, ospreys and other birds of prey.

10.1.1 Nest box birds

One great advantage of nest boxes is that they can be set up and fitted with a camera when unoccupied, with the result that when the boxes are used for nesting there is no disturbance. This is particularly useful with birds such as the barn owl, as these are Schedule 1 birds. The fact that barn owls take readily to nest boxes allows their study without disturbance. Any cameras used should either have no lighting or have infrared LEDs.

Section 2.12.1 describes how to make a camera nest box for small birds. The design can be adapted for larger birds with the help of books such as the *BTO Nestbox Guide* (DuFeu, C. 2005). The main part of any nestbox design will be little changed other than the dimensions, and only the camera section has to be added, either to the top or the side of the box. Some examples are:

- Small birds such as blue tit, great tit, coal tit, sparrow, pied flycatcher, nuthatch.
- Medium-sized birds such as starling, great spotted woodpecker.
- Larger birds such as tawny owl, barn owl, kestrel.

There are non-standard designs of boxes for swift, swallow and house martin.

A standard nest box can also be used with the camera fitted to the side as shown in Figure 10.1. This can be enclosed in a separate box, but if a waterproof camera is used, a

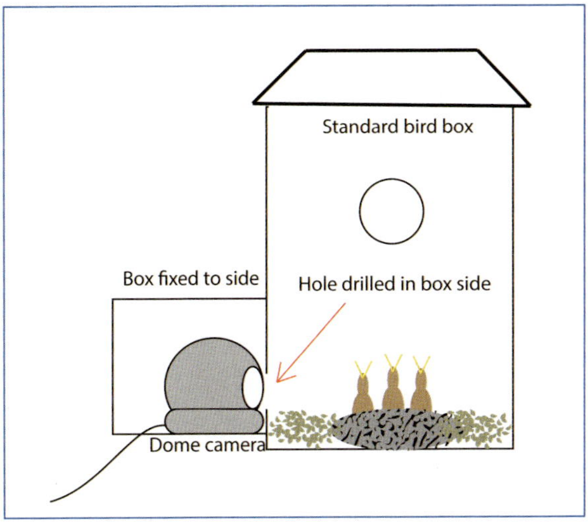

Figure 10.1 Nest box with HD-TVI dome camera on the side.

box may still be useful for greater ease of fixing. For larger boxes a small HD-TVI dome camera or small HD-TVI bullet camera could be used to give HD images.

10.1.2 Ground nesting birds

The siting of cameras near ground nests is something best carried out by experienced nest surveyors, and in many cases may require a licence. The advantage of CCTV is that small unobtrusive cameras can be used. If there is any concern that the very faint red glow from infrared LEDs would cause disturbance, this can be filtered out or cameras with 'dark'

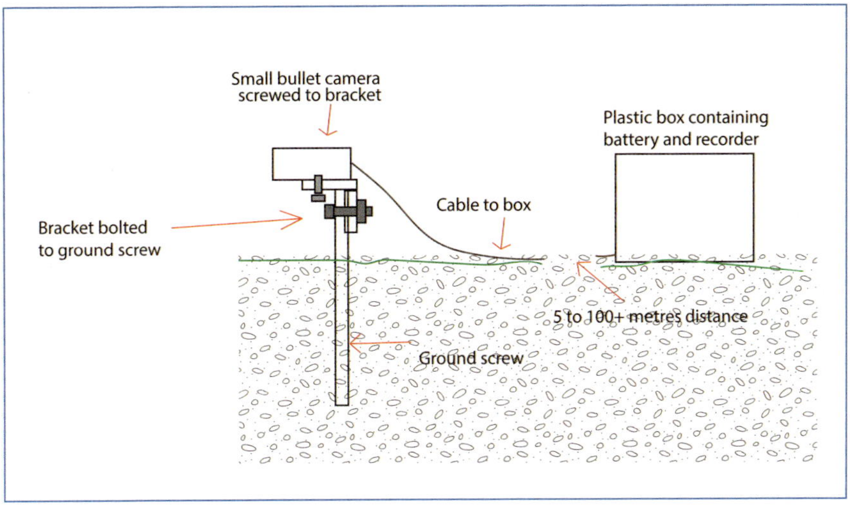

Figure 10.2 Ground nesting bird monitoring setup.

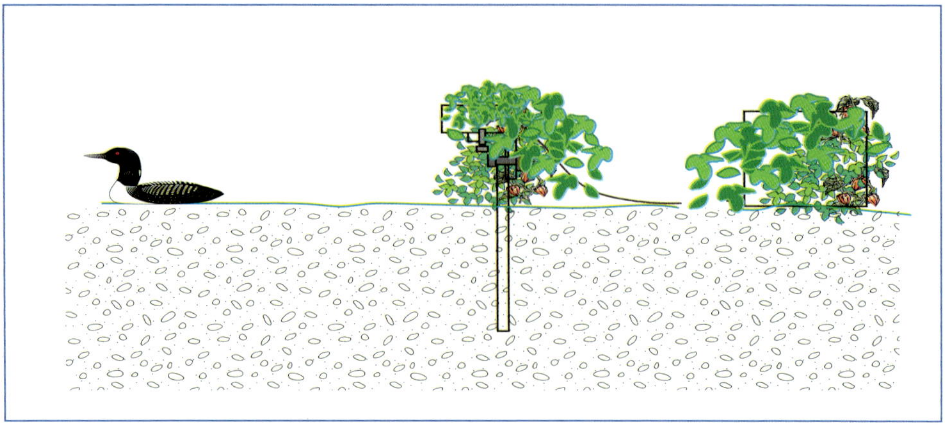

Figure 10.3 Ground nesting bird setup camouflaged.

LEDs of wavelengths around 950 nm can be used. The cameras can be partially hidden in nearby undergrowth or fixed inside a hollow log or artificial rock. They could be fixed to small ground screws to make siting them quick and flexible. The cable, battery and recorder can be contained in a box some distance away, so that changing the battery or the SD card would not disturb the birds.

A simple setup for monitoring ground nesting birds is shown in Figure 10.2 and Figure 10.3.

10.1.3 Tree nesting birds

The equipment for filming tree-nesting birds is similar to any outdoor CCTV monitoring, any difficulty arising from the siting of the camera which would probably be fixed to a nearby tree. It should be stressed that the CCTV components are straightforward; it is the climbing and fixing the camera that requires an experienced or qualified climber.

There is a greater risk of disturbing tree-nesting birds, and a greater likelihood that a licence would be required. Birds such as ospreys and birds of prey that return to known nest sites are less of a problem in some ways as the cameras can be sited before the birds return to nest, but it may still be necessary to obtain a licence, especially in the case of the Schedule 1 osprey. Birds such as the ospreys at Loch Garten in the Cairngorms are already filmed by the RSPB, using IP cameras and live stream video.

Birds such as the heron or buzzard are not Schedule 1 birds, but are still protected from disturbance by the Wildlife and Countryside Act. However, armed with the necessary licences, CCTV filming is feasible, with a portable CCTV system giving the greatest flexibility if the nests are not near to power and/or the internet; otherwise HD cameras could be used.

Not all instances need tree climbing. Young buzzards, for example, spend some time hopping about on trees when they have fledged (see Figure 10.4), and many birds of prey like to perch on poles, and owls perch on branches on the edge of woodland. A ground-mounted CCTV camera close to such places may be successful, as described in section 8.6.

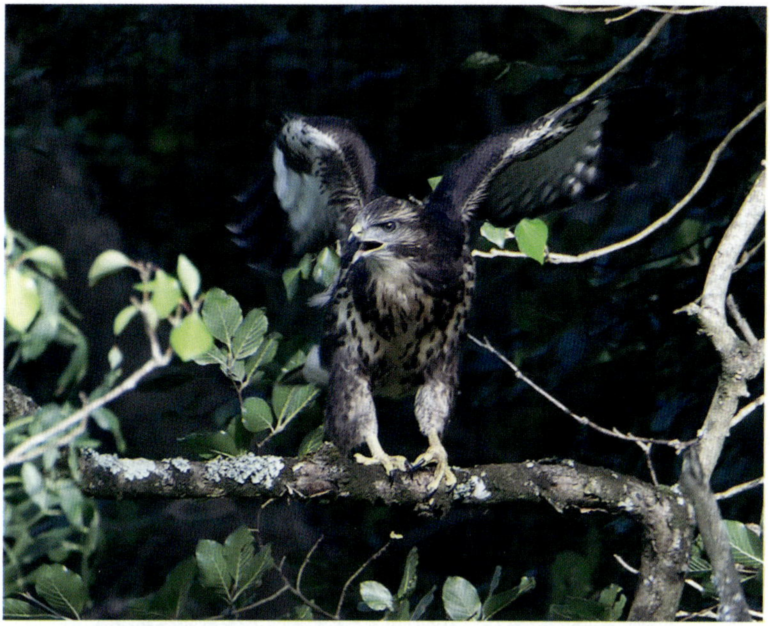

Figure 10.4 Newly fledged buzzard (*Buteo buteo*) tries out its wings.

10.1.3.1 Research

The use of CCTV by researchers in the monitoring of nesting birds is carried out for a range of reasons: monitoring behaviour, determining food varieties fed to young, observation of predation and prevention of wildlife crime. The principle is not new and has been used a great deal in the US, where the equipment used in many earlier projects had few of the advantages of the equipment available today.

Cox *et al.* (2012) found several hundred journal articles describing the use of cameras of various types from 1956 to 2009 in the US. Ratz and Conk (2010) detailed the use of wildlife webcams, in particular commenting on the work of Reif and Tornberg (2006) who identified the issues facing research using CCTV.

Much of the original equipment described in the literature from these reports seems very primitive compared to the equipment available today. Battery technology has advanced, and small, portable DVRs with sophisticated software are now available, with 'ruggedised' versions for difficult environmental conditions.

The basic principles of wildlife CCTV systems do not change for large projects, and although such projects are not within the scope of this book, a knowledge of the basics of CCTV is required before progression to complexity can take place. For large projects, reference to the literature is advised, especially for monitoring outside the temperate climate of the UK, with consideration given to the various CCTV methods described here.

10.2 Reptiles

Reptile monitoring is an ideal candidate for portable CCTV. Reptiles are generally very sensitive to the approach of humans, so it is difficult to monitor behaviour. Surveying tends to rely on spotting the reptile before it spots the surveyor, and with the accent on counting numbers rather than monitoring behaviour.

The equipment setup is likely to be similar to that used for nesting birds, with a suitably long cable from camera to recorder and battery to ensure safety when changing the SD card or the battery. Alternatively a camera could be mounted on a tree or post overlooking a basking area

10.2.1 Adders

Surveying for adders (and other poisonous reptiles) has its hazards, as the health and safety protocol for adder surveys makes clear (ARGUK 2015), but despite the obvious difficulties, it is known that reptiles have favoured basking spots and positioning a camera nearby should pay dividends and minimise risks to the surveyor at the same time. A male adder is shown in Figure 10.5.

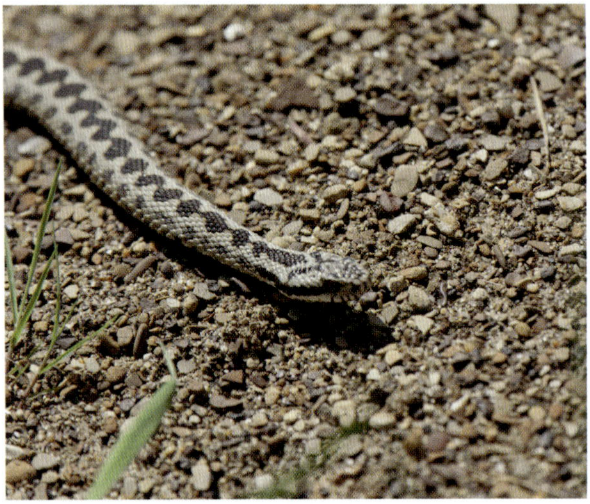

Figure 10.5 Male adder (*Vipera berus*).

The equipment for surveying adders is similar to that used for nesting birds. However, as adders are known to bask on tussocks, the camera should be higher above the ground so that it can see over obstructions such as foliage and tussocks. Although adders may have favoured basking positions, these are not always in the same place, unlike birds on a nest, so the camera must be able to view a greater area. The actual height will depend on the habitat and the choice of ground screw will depend on this.

10.2.2 Grass snakes

Grass snakes (*Natrix natrix*) are known to favour grassy areas near water, especially ponds, and are very sensitive to human approach. Using a similar setup to that used for adders, it should be straightforward to monitor activity near a pond if it is known that grass snakes are present.

10.3 Insects

Insects may seem an unlikely subject for CCTV filming, but a large project studying crickets was conducted by Rodríguez-Muñoz *et al.* (2011). From 2006 to 2008 the team used up to 96 analogue CCTV cameras which were wired to a connection box in the centre of the field being studied. The connection box was connected to a computer 30 m away. The study recorded over 200,000 hours of video, with over 70,000 hours featuring cricket activity.

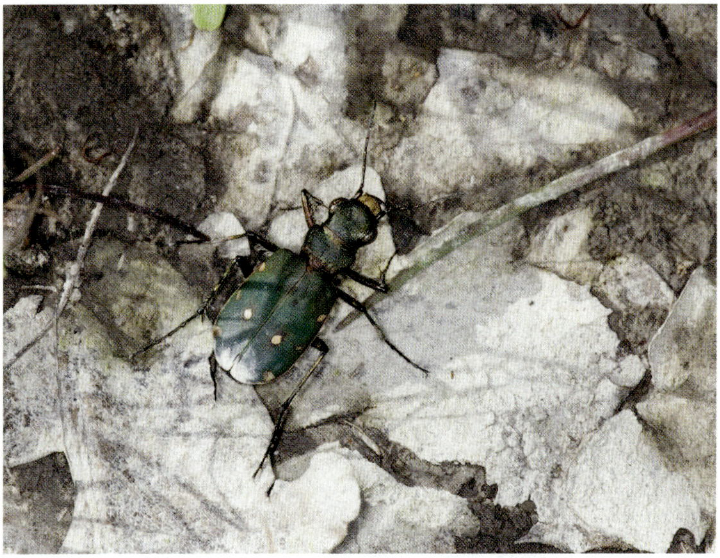

Figure 10.6 Green tiger beetle (*Cicindela campestris*).

Projects of this size, although not within the scope of this book, illustrate the possibilities of using CCTV in complex biological studies. The more sophisticated equipment available today, such as automatic switching from day to night vision, motion-detection recording and off-the-shelf software for monitoring, makes the technical difficulties of large-scale projects less challenging. The problems with analysing large amounts of data remain, but the study confirmed that daily analysis of videos was essential for good management.

For the wildlife enthusiast, CCTV with motion detection recording is an ideal method for studying insects without the disturbance that could be caused by human monitoring. One major advantage of CCTV monitoring is that natural behaviour in the wild is more likely to be observed.

Possible subjects are:

- Butterflies emerging from chrysalises.
- Dragonflies emerging from larvae.
- Nocturnal blue ground beetles.
- Territorial and mating behaviour of dragonflies near ponds.
- Behaviour of ground bees.

As with many wildlife subjects, some prior knowledge of the subject will suggest suitable areas of study and increase the likelihood of success.

10.4 Aquatic mammals

Aquatic mammals in the UK have been the subject of intense conservation activity in recent years. The Eurasian otter (*Lutra lutra*) has come back from near extinction and is now found in every county in England (Crawford 2010), with healthy populations in Wales and Scotland. The water vole (*Arvicola amphibius*) is the subject of re-introduction programmes and, most recently the Eurasian beaver (*Castor fiber)* has been found living wild in Scotland and Devon. The Devon population on the River Otter is now the subject of a Devon Wildlife Trust trial study. The population in Scotland on the River Tay, which probably originated from escapees from private collections, is now large and has been free-living since 2006.

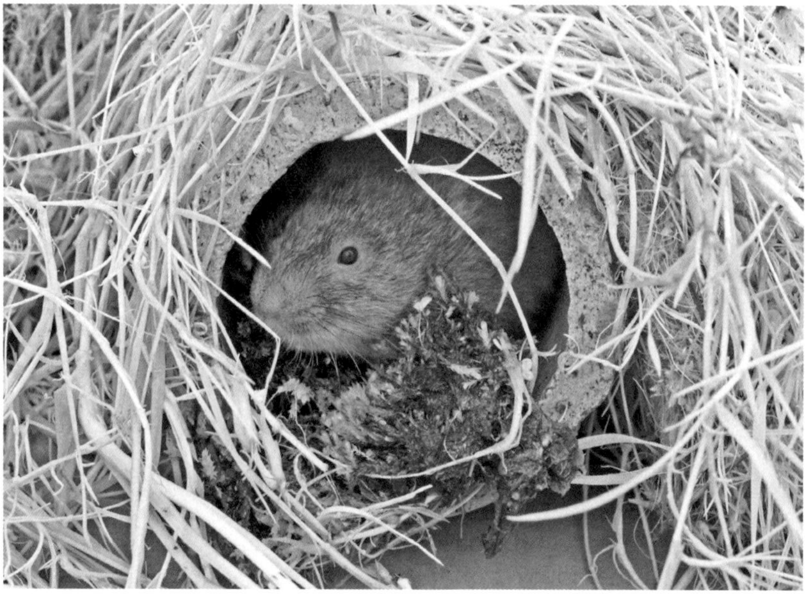

Figure 10.7 Water vole.

They are ideal subjects for portable CCTV using submersible cameras for underwater monitoring, especially in ponds or relatively still water. The robust waterproofing of IP 68 rated bullet cameras would make them suitable for bankside monitoring, where humidity is very high.

Care must be taken to ensure that the animals do not become entangled in cables and that cables are not chewed; therefore cables that can be buried should be used.

11. Comparison of portable CCTV with trail cameras

11.1 Introduction

In order to decide whether your application is best suited to a trail camera or portable CCTV, it is necessary to understand how each works, together with its strengths and weaknesses. This chapter describes trail cameras and compares them with CCTV. Mains-powered CCTV has many advantages such as remote live viewing, greater flexibility in camera type and quality, greater range and greater recording flexibility, so this chapter compares trail cameras with portable CCTV only.

11.1.1 What are trail cameras?

Trail cameras are also known as remote capture cameras, scouting cameras or camera traps. These devices are cameras that are triggered automatically by movement as detected by a motion-detection sensor, and can be pre-programmed to vary the number of images and the time interval between the images. They can take still images, videos or a combination of the two, and can often be used at night as well as during the day.

Unlike most CCTV cameras which film continually, trail cameras are switched on by the motion-detection trigger. Arguably, trail cameras could be described as event monitors as the videos usually have a maximum length of only 60 seconds.

11.1.2 What do they look like?

Figure 11.1 and Figure 11.2 show the main parts of a typical trail camera.

Most trail cameras used today consist of a basic digital camera contained in a box that is water resistant and often has a camouflaged surface. Most have a motion detector mounted on the front which signals the camera to take a picture when triggered. Power is usually supplied by batteries, with the latest models drawing very little power. The images are written to a memory card, usually an SD card, which can be removed to allow the images to be transferred to a computer.

There is also a light sensor that determines when it is dark enough to switch on the LEDs and start infrared (black and white) filming. The better CCTV cameras have a filter to remove infrared light (from the sun) during the day, as this affects colour. At night the filter is removed automatically. Not all trail cameras have this feature.

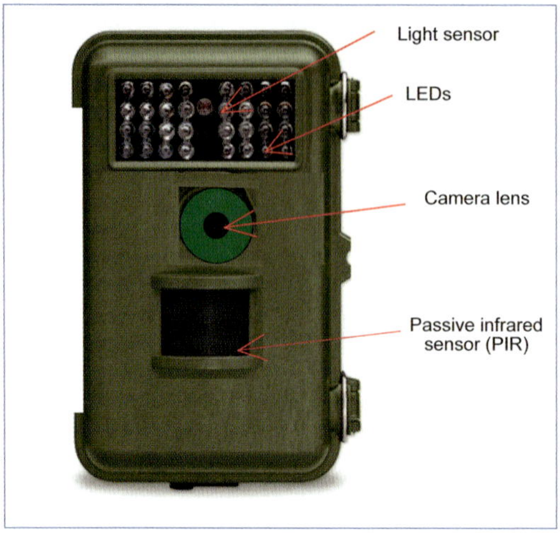

Figure 11.1 Typical trail camera parts.

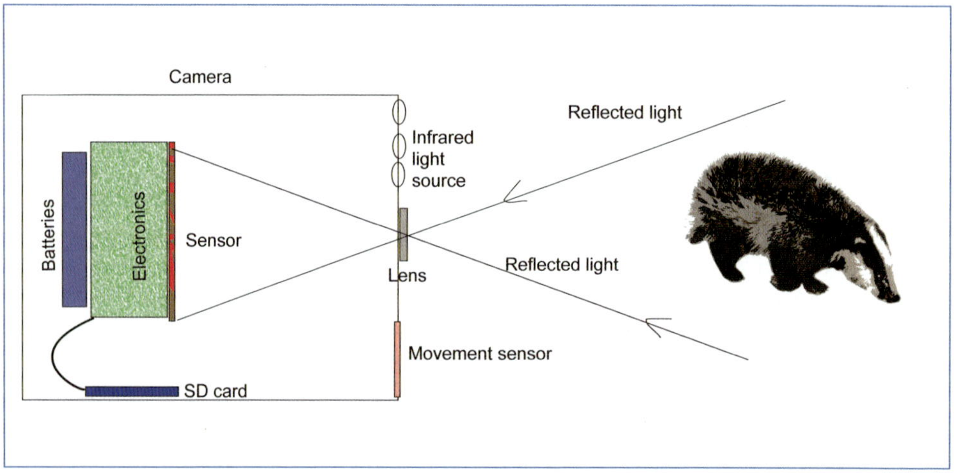

Figure 11.2 Trail camera interior.

11.2 Trail camera details

11.2.1 Structure

11.2.1.1 Camera

The camera within the trail camera casing is a basic digital camera, in most cases with no facility to adjust focal length. In some cases lenses can be screwed onto the front of the camera, but there is less flexibility than with the continuous range of values available with a variable-focus CCTV camera.

11.2.1.2 Movement sensor

Most trail cameras detect motion and trigger the camera using a passive infrared sensor (PIR). This sensor does not detect motion directly, but rather it detects temperature changes. If a warm animal comes within range, the heat given off triggers the camera due to the difference in the heat emitted by a cold background and a warm animal.

Moving branches or leaves can also set off the detector, not directly because they are moving, but because the branches will have absorbed heat from the sun and may be warmer than the surroundings. The moving (warmer) branch will trick the detector into thinking the temperature change has been caused by an animal moving through the detection zone. Moving foliage may also reflect sunlight towards the sensor with the same effect.

In windy conditions there is an advantage in attaching the camera to a post rather than a tree. The tree may be affected by wind and the camera may be triggered by changes in infrared reflections picked up by the sensor or may blur the images due to motion.

The PIR sensitivity can be altered with many cameras. This allows you to adjust the camera to detect larger animals only (lower sensitivity) or small animals (higher sensitivity), and also take account of the ambient temperature. Warm days will need a higher sensitivity than cold days as the temperature difference between the air and a warm animal will be less.

PIR sensors have the disadvantage that it can take a short time for detection and for the camera to be triggered to start filming. This time depends on the camera make, but is often enough to miss events, such as a fast-moving otter or bat, especially if they are not much warmer than the surroundings. Figure 11.3 shows an otter family that nearly 'got away'.

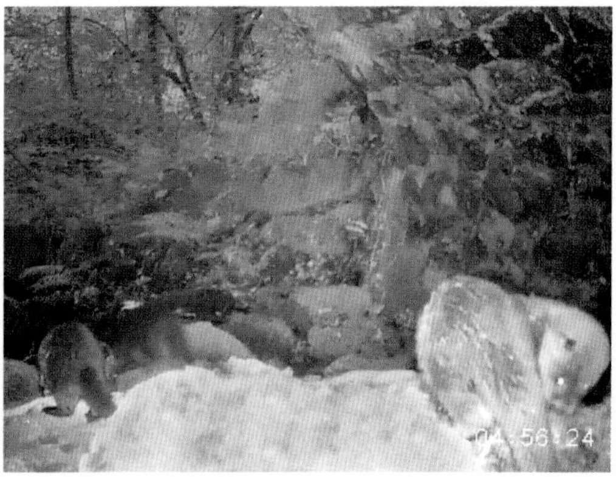

Figure 11.3 This trail camera video of otter and cubs nearly missed the cubs due to the time taken by the camera to detect and trigger recording.

With CCTV filming, motion detection is usually based on actual movement. The recorder detects moving groups of pixels which trigger recording. It is also possible to set up continuous recording with CCTV which is not possible with trail cameras.

11.2.1.3 Power

Power can be supplied by batteries inside the box or from an external source. The latest trail cameras use a number of AA lithium batteries, which can last for several months, depending on the amount of use. Some models allow the use of a rechargeable external source. This is a more expensive option, but is useful for those cameras that use more power. The long-lasting integral power available with trail cameras is a point in their favour as it makes them more convenient than portable CCTV cameras.

11.2.1.4 Recording

The trail camera recorder is integrated into the camera box and can usually take still images or videos, with some cameras also taking HD videos. There is much less flexibility with a trail camera recorder than with a CCTV recorder. Video length can only be adjusted within a very narrow range, with a typical maximum length of 60 s for a trail camera. CCTV recorders, on the other hand, can adjust video length and the actual recording time window can be programmed.

11.2.1.5 Weatherproofing

Most trail cameras consist of a digital camera contained in a weatherproof box. When looking closely at the manufacturers' guides, the term water-resistant rather than waterproof is generally used, so some care is required. The boxes are usually unaffected by drizzle or light rain, but if heavy rain is likely on a regular basis, some protection is advised.

The better CCTV cameras have IP ratings of IP65 to IP68, these designations indicating increasingly waterproof devices. Because trail cameras have to be opened regularly to allow removal of the SD card and to replace the batteries, it is difficult to make them as water resistant as IP-rated CCTV cameras.

11.2.1.6 Cost

Trail cameras can cost under £100 up to £600, and generally speaking, the best images come from the more expensive and reliable cameras. Although CCTV recording needs separate camera, batteries and recorder, the cost for a portable CCTV setup (which is directly comparable with trail cameras) is similar. Figure 11.4 shows a still image from a Reconyx trail camera, generally agreed to be at the top of the quality range.

11.2.2 Usage

11.2.2.1 Day and night

Both trail cameras and CCTV can film in daylight and at night. Many trail cameras can take stills as well as a video, whereas CCTV stills have to be taken from the video, as a rule. In daylight, the better trail cameras can take higher-resolution good-quality still images than the cameras suitable for portable CCTV.

However, at night the difference is less obvious. Illumination at night is supplied by light-emitting diodes (LEDs). These are generally infrared emitting to give 'illumination'

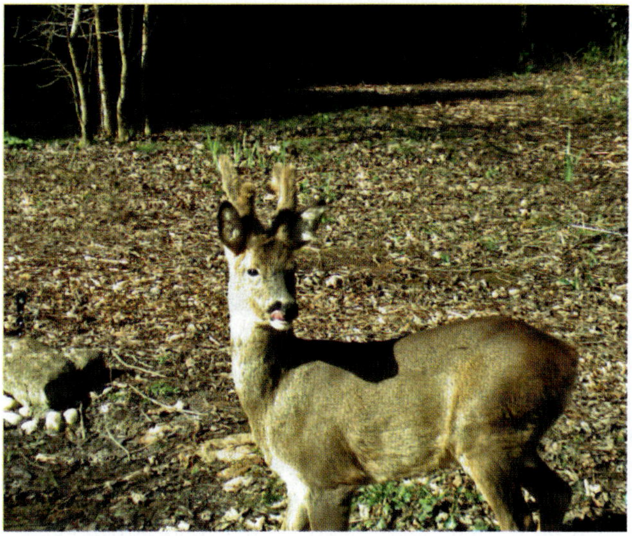

Figure 11.4 Roe buck taken with a Reconyx trail camera.

at night with wavelengths that start in the region of 760–850 nm. The sensors are designed to detect infrared as well as visible light. Most LEDs also emit some red light as the emitted infrared is near the red end of the electromagnetic spectrum and this can be seen as a faint red glow (see Figure 11.5). This is one reason given for wildlife sensing the presence of trail cameras (Meek *et al.* 2014).

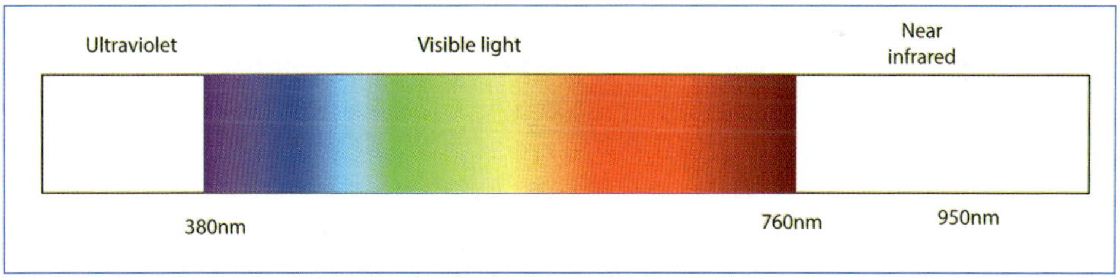

Figure 11.5 The position of infrared in the electromagnetic spectrum.

The latest trail cameras have 'no glow' or black LEDs with wavelengths that start around 950 nm, and these LEDs do not emit any red glow. This effect can be achieved either by LED design so that red is not emitted, or by using a filter to remove the shorter wavelengths. The reasoning behind 'no glow' is that the visible red glow may disturb wildlife, although that has not been my experience. I have also found that the images from 'no glow' can be grainy as there is less infrared light reflected from the target to the sensor.

11.2.2.2 Range

Illumination at night

The working range of the camera varies depending on the conditions. At night, range is the distance that the infrared LEDs can 'light up' the surroundings. This light is not visible to the naked eye, but is detected by the camera sensor, so in this case, range is the distance over which the sensor can detect sufficient light to give a clear image. Beyond this distance, the image becomes dark.

The actual distance varies from camera to camera and depends on the LED intensity and whether or not flash is used, but is typically around 15 m. The PIR detector often has a greater range, at around 20 m. CCTV cameras come in a variety of ranges, starting typically at 10 m and increasing to 50 m or more with more powerful LEDs.

Optimum range

The optimum range will vary from camera to camera, but is often quoted to be around 5 m for trail cameras. As this is well within the detection and lighting up range, this is likely to be based on the lens focal length. Better images will be obtained when the camera points straight at the target. Uneven lighting or focus may be obtained when the camera is at an angle to the target. However, it is impossible to allow for this, as animals do not always walk in the straight lines that suit the camera best, although it is easier if there is a defined trail. In such cases pointing the camera up (or down) the trail is advised.

Effect of temperature on range

Because the PIR detector uses changes in heat levels as its trigger, sensitivity will be reduced in hot weather. Under such conditions, the air temperature will be much closer

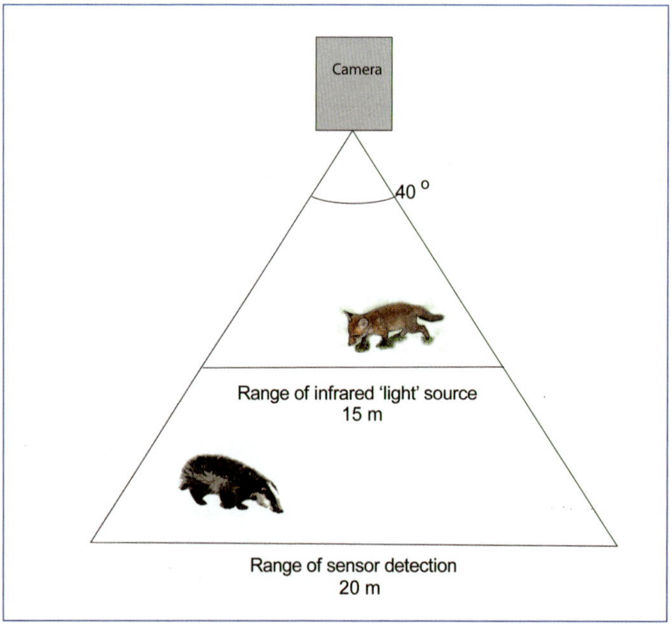

Figure 11.6 With this trail camera, the fox will be detected and will appear on videos. The badger will be detected but there will not be enough light to see it in the video.

to the temperature of the target, and changes will be more difficult to detect. This has the effect of reducing the detection range as the target must be closer for the temperature differences to be detected. On the other hand, in cold weather the difference in temperature between the target and the surroundings will be much higher and the camera will be more sensitive. This has the effect of increasing the range of the detector. At night this can prove a problem as the animal may trigger the sensor while still outside the range of the LED lights (see Figure 11.6). For CCTV cameras, range is entirely dependent on the LED intensity at night.

11.2.2.3 Field of view

The field of view is the area covered by the camera and depends on the angle of view of the lens. The detector may have a different field of view to that of the camera, and ideally should be wider so that the target is centrally placed by the time the camera has been triggered and taken a picture. A typical trail camera angle of view is 30°–40°. For CCTV cameras, the angle will depend on the lens focal length and will vary from 120° to 30°.

11.2.3 Setting up

11.2.3.1 Fixing trail cameras

Tying to a tree

The simplest method for fixing trail cameras is to tie them to a tree. Tapes are usually supplied for this purpose. It is important to ensure that the tape is tight and held securely so that it does not slip. Cable ties can give a secure fixing, but will need cutting off when the camera is to be moved. The trunk should be vertical or very slightly leaning forward, so that the camera points down slightly, as this means that in most cases it will be above the eye-line of the animal. It is important to choose a tree or branch that is big enough that it will not move in wind, as that may cause false triggers or may cause the image to be blurred. It is more difficult to align the camera precisely with this method, especially as ideal mounting posts are not always easy to find.

Tying to a post

In situations where there are no suitable trees, tying the camera to a post is suitable. The post can be round or square in cross-section and should be 5 cm or more in diameter or side. Rough posts are more secure, as the camera can slide down smooth posts, especially if the post is metal rather than wood. The post should first be hammered into the ground until it is stable and then the camera attached. Precise positioning is difficult with this method.

Tripod

Most cameras have a tripod screw and can be attached to a tripod, which makes it easy to position the camera exactly where you want. However, tripods can be knocked over by larger mammals, so this can limit the choice of positions to the edge of a trail rather than in the middle. This method is only suitable for private land where theft is less likely.

Wall/tree screw

If there is a convenient wooden wall, such as on a shed, or a suitable post, branch or tree, an adjustable fixture can be screwed in and the camera attached it. Some fixtures may require that a pilot hole be drilled first, others may simply screw in directly. The advantage of this method is that it is easy to align the camera exactly by adjusting the fixture, and is not dependent on finding a vertical tree of suitable diameter.

11.2.3.2 Fixing CCTV cameras

CCTV cameras usually have to be screwed to a surface, whether a post or tree, although they can be mounted on a tripod using a support. The angle of the camera head can be adjusted up and down, or with some cameras, additionally from side to side. This flexible adjustment makes them much easier to position than trail cameras.

11.2.3.3 Security

Security is a major consideration with cameras of any kind in public spaces. From my experience, CCTV cameras seem less vulnerable because they are screwed to a surface, are more easily camouflaged and people believe they may be filmed. Trail cameras have a large area that must be kept clear so that lens, sensor and PIR are not obscured. This makes them more visible and more difficult to camouflage. Also, thieves look out for trail cameras.

Padlock cable

Provided your trail camera has suitable holes or connections, a cable can be passed through these and secured with a padlock. A thief armed with a cable cutter could still remove the camera, but this method does reduce the odds.

Security box

It is possible to buy a security box that can be fixed to a tree or wall, the trail camera placed inside and the box locked. However, there is still the problem of securing the case, so a cable and padlock will also be required. The security box is of more use in protecting the camera itself from over-curious wildlife and is sometimes called a bear safe.

Security mesh

It is possible to buy security mesh to enclose the camera (Pacsafe). If tightly held round the camera and padlocked into place, this may also prevent removal of the SD card and batteries.

Camouflage

Some trail cameras are produced with a camouflage covering, others are not. In either case, carefully placing leaves or branches round the camera can help to camouflage it. Care must be taken not to cover the PIR sensor or the lens.

Height

A camera placed high in a tree, looking down, will be difficult to remove, and this is one option. However, it will also make it more difficult to check the SD card, as a ladder may be required.

Other methods

Although it will not protect the camera from theft, using a cheap camera to watch the trail camera is a method of recording the theft and hopefully making it easier to identify the thief.

'No glow' cameras are impossible to see at night as there is no tell-tale red glow, so this is one option.

Prevention is still the best method, and siting and camouflaging the camera carefully so that it is difficult to see may be the best method of all.

11.2.4 Working with images

11.2.4.1 Recording images

Most trail cameras store images on an SD card, and it is essential to read the manual, as not all makes of camera work with the newest SD cards. Most cameras come with an SD card but it may only store less than 1 GB. If the camera is to be kept in place for several days, and especially if videos are taken, a higher-capacity card may be required. Again, care is needed as most cameras have an upper limit of storage. Some may support cards up to 32 GB, but in many cases 4 or 8 GB is more than enough capacity.

The mobile recorders used for portable CCTV also use an SD card, with 32 GB giving sufficient capacity for several days, depending on the number of recordings made.

11.2.4.2 Viewing images

Most trail cameras require the SD card to be removed and the videos downloaded to a PC for editing using video-editing software. Mobile recorders for portable CCTV work in the same way except that CCTV recorders often provide software for rough editing.

11.2.4.3 Remote viewing

Wireless

Some trail cameras come with an additional wireless module, allowing transmission of images (not videos) to PC, tablet or smartphone. They only have a range of about 100 m, which may be of limited use, depending on your situation, but this may be a useful feature for home viewing at night or in bad weather. This system is similar to wireless CCTV.

Mobile technology

Some of the more expensive trail camera models come with a SIM card and make use of 3G mobile phone technology, so that still images (not videos) can be sent to a mobile phone or via email. Such cameras may come with a solar panel to extend battery life in the transmitting unit. There is an additional cost for transmission, usually to a mobile network company, and the system will only work if there is a mobile signal where the camera is placed.

With CCTV, wireless cameras using 5.8 GHz can transmit signals (including video) several kilometres as described in section 2.13.1. Also, some CCTV recorders can transmit videos live using 3G, although at low resolution. 4G systems are being developed.

11.2.4.4 Image quality

Sensor resolution

Sensor resolution is a term well known to digital photographers and refers to the number of pixels or individual light-gathering cells on the camera sensor; this resolution is typically quoted in megapixels. It is these cells that convert light to an electrical signal which in turn sends data to the memory card. For trail cameras, resolution can vary from 3 to 12 megapixels, but this refers to still images only. Video resolution is typically 1,280 × 720, which is just under 1 megapixel.

The CCTV cameras most suitable for portable systems require a low current as used by lower-resolution analogue cameras, which have resolutions closer to 0.5 megapixels. However, as with standard digital cameras, the number of pixels is only part of the story. The size and quality of the pixels together with the quality of the camera electronics is as important if not more so. Thus although CCTV videos from 0.5 megapixel cameras have a smaller screen size than a trail camera with an HD video resolution of 1 megapixel, CCTV video quality can be as good or even better, especially at night.

Light levels

Light levels have a great deal of effect on image quality. During the day, higher (natural) light levels will produce less noise and a higher-quality still or video image. Trail camera images in woodland will tend to be of lower quality due to lower light levels. CCTV cameras, on the other hand, are sensitive to low light levels and can produce good-quality images in woodland.

Importance of image quality

When trail cameras are used to highlight the presence of wildlife as an aid to making digital camera wildlife photography more efficient, trail camera image quality may not matter too much, and in such cases a cheaper camera may suffice. However, trail cameras are being used increasingly for survey purposes and to produce videos, and in this case image quality is important.

Good-quality CCTV cameras have advanced electronics to improve image quality. Noise reduction, sharpening and contrast control are just a few of the advantages that CCTV cameras offer. Some also have 'smart IR' whereby overexposure of nearby objects by the LEDs is avoided.

11.3 Summary

Trail cameras and portable CCTV each have their strengths and weaknesses. Trail cameras are very convenient, but very vulnerable to theft. Portable CCTV is much less convenient but is more flexible and able to track fast-moving subjects. It is only when both have been tested and the subject matter studied that an informed choice can be made for a particular portable application.

Further details on trail cameras can be found in the publication by Rovero and Zimmermann (2016).

12. Comparison of CCTV with remote triggered DSLR

12.1 Introduction

Camera traps are not a new idea, and at one time, remotely triggered film or digital cameras were used as camera traps, with the breaking of an infrared beam as the trigger device. The all-in-one trail camera is the modern-day successor to these devices. However, trail cameras and CCTV cameras do not always give high-quality images, or those that do, give images that are of good quality, but are of a lower pixel count. For the highest quality images/videos, a high-quality digital camera (DSLR) is required. Electronic trigger devices compatible with such cameras are now available, and these systems can yield high-quality images. A typical triggered system is shown in outline in Figure 12.1.

This is a much more expensive method, as cameras, lenses and triggering devices are costly and sites may be limited by the necessity to protect the cameras from theft – a DSLR camera and trigger equipment being all too obvious. For this reason, such methods are best used on private land or at a secure site, or else when the photographer is close by, but far enough away so that the wildlife is not disturbed.

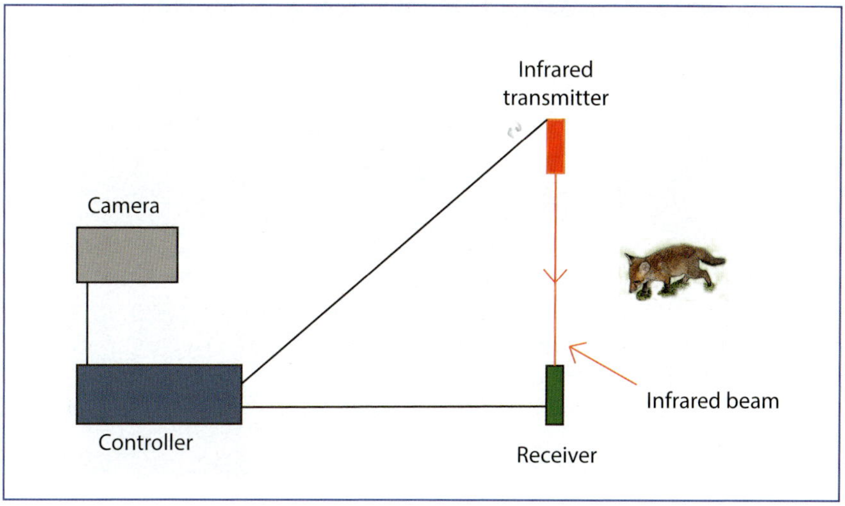

Figure 12.1 Basic infrared trigger system. When the beam is broken the controller fires. If the transmitter has a battery it is not connected to the controller.

For the purposes of describing how to use the equipment, I will predominantly use the example of high-speed flash for capturing birds in flight, but there are many different applications, and each application may use the equipment in a slightly different way.

This is a truly mobile method as all the equipment is run by batteries, although repeated triggering may restrict working time to a few hours.

12.1.1 What are triggered cameras?

Trail cameras are triggered to take pictures by the integrated PIR motion detector, and CCTV cameras film continuously with recording being triggered by the recorder or PC software. Triggered DSLR cameras use an external triggering device to fire the camera shutter. They can be triggered by sound, light, by breaking/making an infrared beam or by pressure switches. The sensing devices are connected to an electronic controller. This is the most expensive form of remote photography, as the cost of controller, sensors and camera can be high, or very high, depending on the method used.

12.1.2 What do they look like?

The trigger system consists of an infrared transmitter and receiver usually contained in small boxes connected to stands, and a controller in a larger box. Controllers allow variation in trigger sensitivity, and the time before the system is reset. One make of controller is shown in Figure 12.2.

12.1.2.1 Weatherproofing

Generally this method is best used for short sessions when the photographer is not far away, and for this reason choosing a day with dry weather is an option that avoids

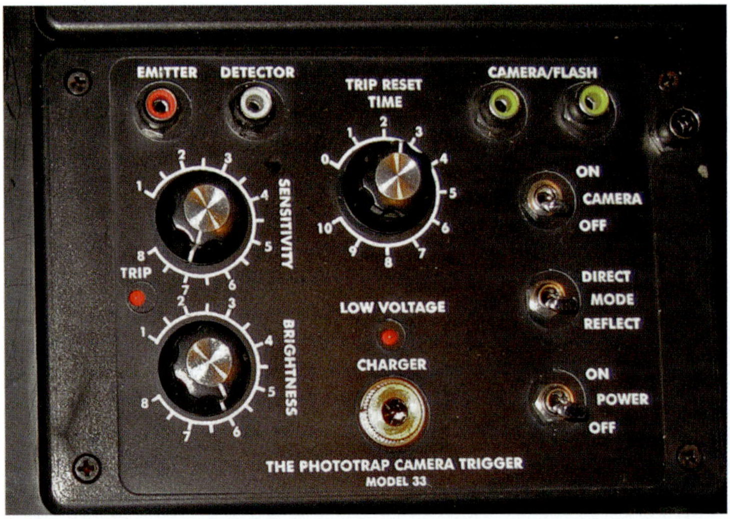

Figure 12.2 One form of controller. Both transmitter (emitter) and receiver (detector) are plugged into the controller. The camera is connected to one yellow socket and the rightmost black socket.

weather problems. If you can leave your equipment outside, I have found a portable hide to be very useful for positioning camera and tripod, together with the triggering equipment.

12.1.2.2 Power

Both camera and trigger equipment are run on batteries. Unlike trail cameras and CCTV, triggered DSLR shooting is generally used for shorter time intervals and batteries are sufficient. This is not generally a technique in which equipment is left for long intervals.

12.1.2.3 Cost

DSLR cameras vary in cost, but are usually over £500 and up to several thousands of pounds. Triggering equipment varies from a few hundred to a few thousand pounds, depending on the brand and accessories used.

12.1.3 Examples of use

By day:

- Beam broken by the subject.
- Reflecting of the beam to the receiver by the subject (the easiest method to set up).
- Crossed beams giving greater focusing accuracy.
- High-speed flash for fast-moving subjects.

By night:

- Beam broken/reflected white light flash where the wildlife will not be affected by flash.
- Beam broken/reflected Infrared flash where wildlife is affected by flash. A camera sensitive to infrared must be used.

12.2 Triggered camera system details

12.2.1 Structure

12.2.1.1 Camera

Any camera which allows remote shutter release can be used. Digital cameras today generally have the correct fitting and require a camera-to-controller cable specific for the make and model. Most trigger systems have cables for the main brands of camera.

12.2.1.2 Trigger

The infrared transmitters and receivers are small devices that can be attached to small stands. I have found that small ground screws are excellent for this purpose. Some devices have internal batteries, others are powered by the controller.

12.2.1.3 Power

The system is entirely powered by batteries, which vary in type according to the model used. The Phototrap in Figure 12.2 has a rechargeable battery in the controller.

12.2.1.4 Recording

Video-recording is not reliably triggered by the system, and its main use is to supply high-quality still images, stored on the camera card.

12.2.1.5 Weatherproofing

The controllers are not waterproof and must be protected from wet. The transmitter and receiver may be weather resistant, but are not usually waterproof.

12.2.1.6 Cost

The trigger system cost varies according to manufacturer from £300 to £2,000.

Battery life
This depends very much on the system used and how frequently it is triggered. I have used the system for half days at a time and have found that camera batteries and those in the triggering equipment last for several sessions. However, leaving a system out all night has drained the camera battery.

12.2.1.7 Image quality

Sensor resolution
One of the main reasons for using triggered DSLR cameras is that the images have much higher resolution than trail cameras or CCTV cameras. DSLR cameras have sensors with a resolution of upwards of 10 megapixels. Some have a resolution of 50 megapixels. They also have high-quality electronics and internal software to enhance the image.

Light levels
For daylight shooting, especially if flash is used, light levels will not be a problem, as the camera ISO can be increased. Night shooting is a different matter, as flash is required. There are differences of opinion as to whether flash at night is ethical as the wildlife subject may be disturbed. One solution is to use floodlights. I have found that badgers and foxes are not disturbed by a fixed light source, it is the suddenness of flash that can be a problem to them, and so floodlights offer an acceptable light source, provided that a power source is close at hand.

12.2.2 Usage

To justify the cost, trigger systems are best used for situations that are difficult to manage with either trail cameras or CCTV, or to complement them. Triggered systems are used when there is a need:

- To use high-quality cameras and thus obtain higher-resolution images.
- To capture high-speed motion of wildlife.
- To photograph birds in flight.
- To photograph insects in flight.

Although the method of taking an image is different from CCTV, there are parallels. The trigger system is analogous to motion-detection recording, except that the 'recording' is a still image on the camera card.

12.2.3 Triggering method

Breaking the beam

The commonest triggering method is to interrupt an infrared beam so that the receiver does not detect it. This causes the controller to send a pulse to the camera to fire the shutter.

In Figure 12.3 an infrared (or laser) beam from the transmitter is detected by the receiver. The controller does not send a signal to the camera and no photograph is taken.

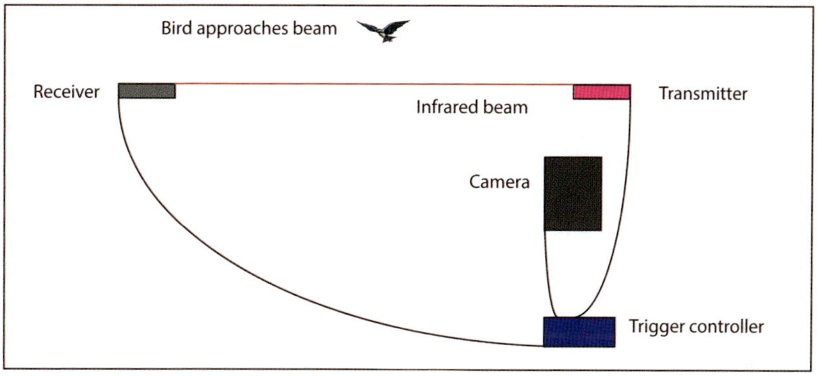

Figure 12.3 Trigger system before the trigger.

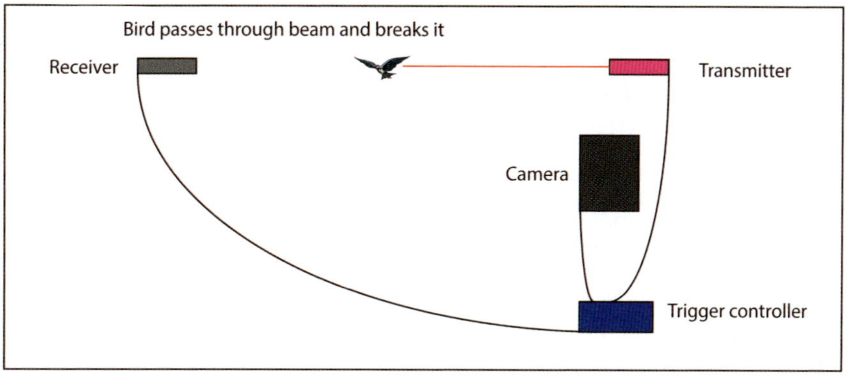

Figure 12.4 Trigger system after the trigger. The beam cannot reach the receiver and the camera takes a picture.

In Figure 12.4 the bird passes through the beam, it blocks the receiver and 'breaks' the beam. The controller detects this and sends a signal to the camera, which takes a photograph.

The system can be setup in different ways:

- The beam can be reflected to the receiver by the subject so that *making* the beam takes the photograph. This is known as the reflect position.
- Two sets of detectors with crossed infrared beams can pinpoint more precisely the subject's position when the beam is broken.

12.2.4 Uses of triggered cameras

12.2.4.1 Range

A triggered DSLR has a range dictated by the length of the cables connecting the camera to the sensors and controller, and can be from a few centimetres to 4 m, depending on the application. The focal length of the lens on the camera also affects the range. For example, with a short focal length lens, the subject must be close to the camera.

12.2.4.2 Field of view

Field of view is dictated by the lens used, and the size of the camera sensor. In most cases this is the same as the view through the viewfinder. It is important to establish the field of view of the camera/lens combination, as different applications will require different framing. Photographing insects for example will make use of a much smaller field of view than a bird in flight. It is important to take practice shots to ensure you have the correct field of view and have captured as much (or as little) of the scene as you want.

12.2.4.3 System trigger speed

For slow-moving wildlife, the speed at which the system responds is relatively unimportant, but if you want sharp images of fast-moving objects, speed is very important, and high-speed flash essential.

12.2.4.4 High-speed images

There are several ways to obtain high shutter speeds and/or high-speed flash: a camera that has high shutter speeds; high-speed sync selected on the flash gun; high-speed flash obtained by reducing power; external high-speed shutter. These are detailed below:

Fast camera shutter speed

Some top-end cameras have a shutter speed of up to 1/8,000 s, which can be used without flash in well-lit situations. In many situations, however, there will be insufficient light for a well-exposed subject, especially if a small aperture is used. Increasing ISO helps, but usually adds enough noise to render the image less than sharp. For many situations, photography of fast-moving wildlife requires flash.

Flash with high-speed sync

The problem with using flash is that flash and shutter have to be in sync otherwise the shutter curtain will cover part of the image. For this reason, the highest shutter speed for normal flash work is around 1/250 s. Thus, although the flash duration may be short (e.g. 1/1,000 s), ambient light will continue to enter while the camera shutter is open and this is likely to cause blurring of the image.

With high-speed sync, the shutter and flash work together to give faster shutter speeds in sync with the flash gun which prevents the shutter curtain from obscuring part of the image. High-speed sync can be setup on many camera/flash combinations, provided that the flash gun has this feature. The disadvantage is that the subject has to be quite close as the power and hence the range of the flash falls off quickly as the sync speed increases.

The image in Figure 12.5 was taken using flash and high-speed sync. The shutter speed was 1/6,400, aperture was f/11 and ISO was 640. Additional flash guns can be used to light the background if required.

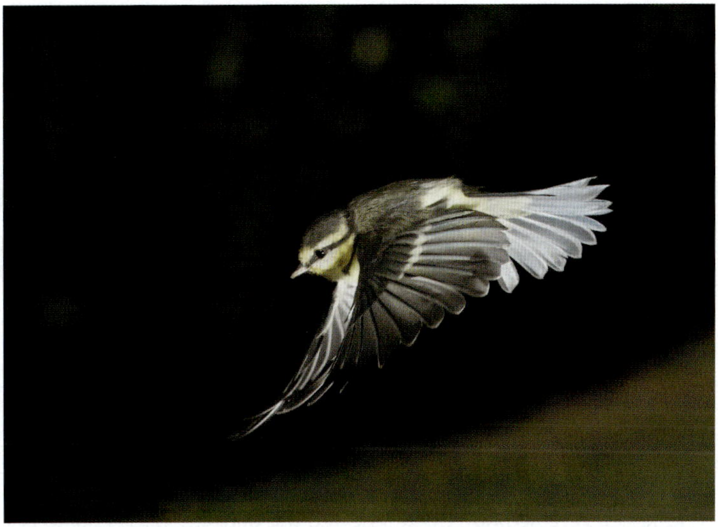

Figure 12.5 A blue tit photographed with high-speed flash triggered by a controller.

External high-speed shutter

The very best high-speed flash images, shot with a small aperture and a very high shutter speed for sharpness, are achieved with an external shutter. This also reduces the problem of shutter delay or lag. A flash may fire as soon as triggered, but a camera will take several milliseconds to react and open its shutter. This can result in fast-moving subjects being in the 'wrong' place and either out of shot or out of the focal plane. The solution may be an external shutter.

An external high-speed shutter is attached to the camera lens and is triggered by the controller. The camera is used in bulb (shutter open) mode so there is no camera delay. Bulb mode can be used in daylight because, although the camera shutter is open, the external shutter will not open until triggered. Light will only reach the sensor when the external shutter is open.

This is not a cheap solution but is one that will give the best results for fast-moving wildlife, especially small subjects such as insects.

12.2.4.5 *Fast-moving subjects*

Correct positioning of sensors and camera is essential, but can only be achieved with practice. There are several issues to consider: ensuring the subject is photographed while in focus; sensors positioned correctly; flash triggered at the correct point in the target's motion.

Birds

In Figure 12.6, food was placed on a log, and the sensors (transmitter and receiver) were placed in the reflect position on a tripod between the food and a perch known to be favoured by the birds. After watching for some time, a common flightpath from perch to food was determined and the sensor position adjusted.

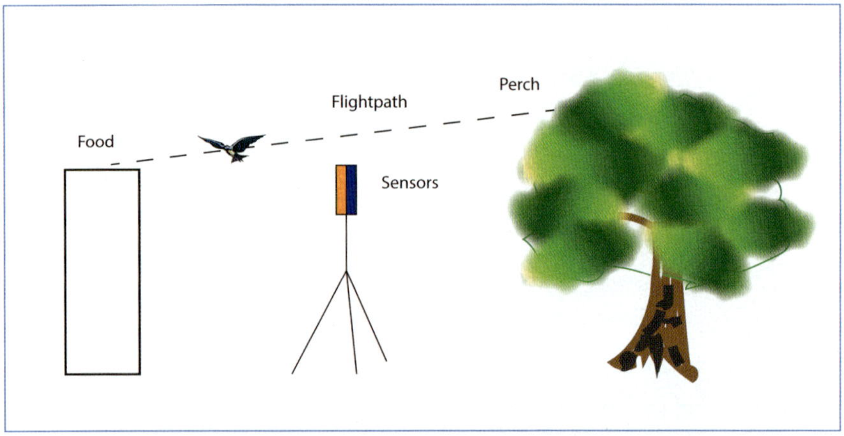

Figure 12.6 The sensors are positioned after noting the path the birds take to get to the food.

As the birds fly towards the food, they pass over the sensors, reflect the beam to the receiver and trigger the camera. Only a proportion of shots will have the bird in the plane of the flightpath, as it may be behind the estimated position or in front of it. This is an advantage if flash is used, as it ensures that the flash is fired in, say, only one in five shots. As flash does disturb the birds to some extent, this ensures that for most passes they are not disturbed. In addition, the setup should only be in place for an hour at most, as disturbance will then be reduced to a minimum.

12.2.4.6 *Larger subjects*

One of the problems with positioning sensors for larger mammals is that the sensors have to be firmly fixed. The most convenient method is to use a tripod, but tripods can be knocked over, are large and may have a perceivable amount of human scent attached.

Furthermore, setting up tripods for sensors, as well as camera and possibly flashes, is likely to make the site rather busy.

The sensors in Figure 12.7 are bolted to ground screws. The ground screws can be hammered or screwed into place and unscrewed out of the ground afterwards. They are cheap, stable and ideal for sensors.

Once an animal crosses and breaks the beam, the camera is triggered by the controller.

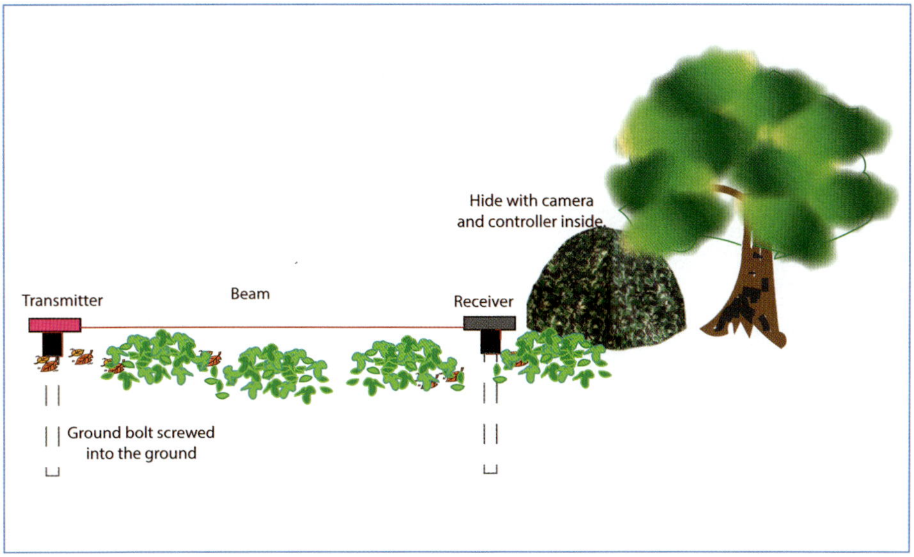

Figure 12.7 Trigger setup for larger wildlife.

12.3 Summary

Triggered photography has the same basic elements as CCTV: camera, power (batteries) and recorder (the camera card). The trigger equipment takes the place of the motion detector.

The execution of triggered photography is very different from CCTV as only stills can be obtained reliably (some cameras can be triggered to start a video, but it is more difficult to stop the video). However, there is often a need for high-quality images to supplement CCTV videos, and the modern triggered system could be described as a desirable accessory to increase the usefulness of CCTV, which has the potential to revolutionise some aspects of wildlife monitoring.

Appendix

CCTV and trail camera comparison

Table A1 Comparison of CCTV and trail cameras

Feature	Analogue CCTV	HD-TVI CCTV	IP CCTV	Trail camera
Still images	From video	From video	From video	✔
HD video	✖	✔	✔	✔
Mains power required	Not essential	✔	✔	✖
Battery powered	✔	High-capacity battery only	✖	✔
Continual filming	✔	✔	✔	✖
Heat sensing motion detection	✖	✖	✖	✔
Motion sensing detection	✔	✔	✔	✖
Scheduled recording	✔	✔	✔	✖
Pre-record	✔	✔	✔	✖
Flexible post-record time	✔	✔	✔	✖
Masking on DVR or with PC software	✔ Not all portable DVRs	✔	✔	✖
Uses a DVR for recording	✔	✔	✖	✖
Uses PC software for recording	With capture device	✖	✔	✖
Submersible cameras	✔	✖	✔	✖
Variable focus cameras	✔	✔	✔	✖
Variable LED range	✔	✔	✔	✖
Suitable for portable CCTV	✔	With high-capacity battery	✖	NA
Wide range of camera types available	✔	✔	✔	✖
Wide range of recorder types	✔	✖	NA	✖
Cameras with noise reduction	✔	✔	✔	✖
Easy to use	✖	✖	✖	√
Remote viewing	✔	✔	✔	✖
Image quality	Good	High	High	Good

Equipment used

This is a list of the equipment I have used. It is by no means exhaustive, and there are many other brands and models that would be satisfactory.

Cameras

Genie ZB2812IR analogue bullet camera. 600 TVL, variable focus lens (2.8–12 mm), 250 mA, 12 V DC, range of LEDs 20 m.

Genie ZD2812IR analogue bullet camera. 1,000 TVL, variable focus lens (2.8–12 mm), 700 mA, 12 V DC, range of LEDs 30m.

Huviron SK-P461/HT21AIP HD-TVI camera. 2.2 megapixels, variable focus lens (2.8–12 mm), 650 mA, 12 V DC.

Vivotek IP8152 IP mini camera. 1.3 megapixels, variable focus lens (3.3–12 mm), no LEDs.

Vivotek IB8367 IP bullet camera. 2 megapixels, variable focus lens (2.8–12 mm), range of LEDs 30m.

Recorders

Huviron HD-TVI DVR 1080p Hybrid. Four-channel input for HD-TVI and analogue cameras. 12 V DC.

Genie CCTV SD-DVR. 12 V DC, 150 mA, motion-detection recording, pre-record.

Batteries

Tracer lithium polymer batteries. 8 and 22 Ah.

Tracer 12 V 80 Ah LiFePO4 Peli™ Case Kit.

Numax 110 Ah deep-cycle sealed lead-acid battery.

Equipment suppliers

The equipment and manufacturers listed here are those I have used – there are many more I have not yet tried.

Batteries

http://www.tracerpower.com/ (Lithium batteries)

https://advancedbatterysupplies.co.uk/ (Lead acid batteries)

Bird box camera suppliers

http://www.handykam.com/

http://www.wildlife-cameras.com/

CCTV camera and recorder suppliers

https://rfconcepts.co.uk/
http://www.onlinesecurityproducts.co.uk/
https://www.spycameracctv.com/
http://www.icode.co.uk/icatcher/

CCTV camera and recorder manufacturers

http://geniecctv.com/ (analogue cameras and SD-DVR recorder)
http://www.vivotek.com/ (IP cameras)
http://www.huviron.com/ (HD-TVI cameras and recorders)
http://www.ovation.co.uk/ (ruggedised DVRs)

CCTV components

http://www.maplin.co.uk/ (cables, solar panels)
http://uk.rs-online.com/web/
http://www.rapidonline.com/
http://www.solwise.co.uk/ (home plugs)

Miscellaneous

http://www.peli.com/ (Peli waterproof cases)
https://www.ledhut.co.uk/ (LED strip lights)
http://www.reallyusefulproducts.co.uk/uk/ (large plastic boxes)
http://www.jackpyke.co.uk/ (camouflage net)

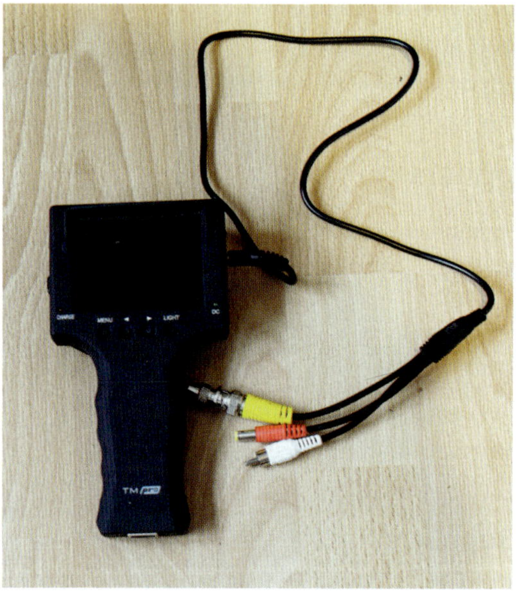

Figure A1 Test monitor.

http://www.groundbolt.co.uk/ (ground screws)
http://www.wildlifeacoustics.com/ (Bat detectors/recorders)
http://bushnell.eu/uk/produits/all/trail-cameras/natureview-cam-hd/ (trail cameras)
http://www.reconyx.com/ (trail cameras)
http://www.phototrap.com/index.htm (DSLR triggering)
http://www.flaghead.co.uk/pages/triggersmart/triggersmart-intro.html (DSLR triggering)

Monitors

https://www.tlc-direct.co.uk/Products/TSTMPRO.html (test monitor). An example test monitor is shown in Figure A1.
http://www.lilliputuk.com/ (small monitors).

Software

http://www.icode.co.uk/icatcher/ (CCTV camera to PC display management)
Adobe Premiere pro (video-editing)

References

ARGUK (2015) *Make the Adder Count. Amphibian and Reptile Groups of the UK.* Accessed at: http://www.arguk.org/mtac-2014 (7 November 2015).

Beaumont, W. (2002) *Guidelines for Electric Fishing Best Practice.* Environment Agency Report W2-054/TR. Accessed at: https://www.gov.uk/government/uploads/system/uploads/attachment_data/file/290344/sw2-054-tr-e-e.pdf (27 October 2015).

Cox, WA. et al. (2012) *Development of Camera Technology for Monitoring Nests.* University of Nebraska. Accessed at: http://digitalcommons.unl.edu/usgsnpwrc/250/ (20 October 2015)

Crawford, A. (2010) *Fifth Otter Survey of England 2009–2010.* Environment Agency Technical Report. Accessed at: http://www.thepredationactiongroup.co.uk/images/EA_otter_survey_oct10.pdf (17 November 2015).

Davies, J. (ed) (2001) *Marine Monitoring Handbook.* Joint Nature Conservation Committee. Accessed at: http://jncc.defra.gov.uk/MarineMonitoringHandbook (27 October 2015).

DuFeu, C. (2005) *Nest Boxes.* British Trust for Ornithology. Accessed at: http://www.bto.org/sites/default/files/u15/downloads/publications/guides/nestbox.pdf (30 October 2015).

Eltringham fishings on the Tyne. Environment Agency fish counter at Riding Mill. Accessed at http://www.fishtyneriver.com/fishcounter.php (22 November 2015).

Health and Safety Executive. *Using Storage Batteries Safely.* Accessed at: http://www.hse.gov.uk/pubns/indg139.pdf (20 February 2016).

Humphreys, E. *et al.* (2010) *An Examination of Reptile and Amphibian Populations in Gardens, the Factors Influencing Garden Use and the Role of a 'Citizen Science' Approach for Monitoring their Populations within this Habitat.* BTO Research Report No. 572. Accessed at: http://www.bto.org/sites/default/files/shared_documents/publications/research-reports/2010/rr572.pdf (27 October 2010).

Meek, P.D. *et al.* (2014) Camera traps can be heard and seen by animals. *PLOS ONE* 9(10): 0110832. Accessed at http://journals.plos.org/plosone/article?id=10.1371/journal.pone.0110832 (29 October 2015).

Ratz, J.M. and Conk, S.J. (2010) *Use of Wildlife Webcams – Literature Review and Annotated Bibliography.* US Geological Survey Open-File Report 2010-1306, p. 42. Accessed at: http://pubs.usgs.gov/of/2010/1306/pdf/OF10-1306.pdf (20 October 2015).

Reif, V. and Tornberg, R. (2006) Using time-lapse digital video recording for a nesting study of birds of prey. *European Journal of Wildlife Research* 52(4): 251–258. Accessed at: http://www.researchgate.net/publication/225340101_Using_time-lapse_digital_video_recording_for_a_nesting_study_of_birds_of_prey (29 October 2015).

Rodríguez-Muñoz, R., Bretman, A. and Tregenza, T. (2011) Guarding males protect females from predation in a wild insect. *Current Biology* 21: 1716–1719. Accessed at: http://www.selfishgene.org/Tom/Papers/RRMetal_CB11.pdf (29 October 2015).

Rovero, F. and Zimmerman, F. (2016) *Camera Trapping for Wildlife Research.* Exeter: Pelagic Publishing.

Russo, D. *et al.* (2004) Roost selection by barbastelle bats (*Barbastella barbastellus*, Chiroptera: Vespertilionidae) in beech woodlands of central Italy: consequences for conservation.

Biological Conservation 117: 73–81. Accessed at: http://onlinelibrary.wiley.com/ (20 October 2015).

Selbert, A. *et al.* (2015) Bidirectional echolocation in the bat *Barbastella barbastellus:* different signals of low source level are emitted upward through the nose and downward through the mouth. *PLOS ONE* 10(9): 0135590. Accessed at: http://journals.plos.org/plosone/article?id=10.1371/journal.pone.0135590 (29 October 2015).

Washburn, E. *et al.* (2008) *Using Video Images for Fisheries Monitoring*. Environment Agency Science Report SC050022/SR2. Accessed at: https://www.gov.uk/government/uploads/system/uploads/attachment_data/file/290278/scho0408bnyb-e-e.pdf (27 October 2015).

Wildlife and Countryside Act 1981. Accessed at: http://www.legislation.gov.uk/ukpga/1981/69 (29 October 2015).

Index

Page numbers in *italic* indicate figures.